WINONA AND THE KING PHILIP LOCATIONS

942 Ninth Street
Lake Linden, Michigan 49945

ISBN: 0-942363-43-4

Forty-fourth of a Local History Series

Printed in the United States of America by Welden H. Curtin, Lake Linden, Michigan

AUTHOR'S NOTE

The data from which this book was compiled were secured through numerous sources. From the many conflicting statements which appear in the original manuscripts and newspapers, this author has chosen the data and incidents which appealed to his reason, after all possible research, as being most accurate. If any errors are detected or important information missing, please contact me. These entries are a combination of many sources, most of which are listed at the end of this compilation.

Sincerely,

Clarence J. Monette
Author

INDEX

Winona, a hamlet of Elm River Township, Houghton County, is located approximately halfway between Ontonagon and Houghton, and about one and a half miles off M-26. According to long-time residents, Winona is pronounced Why-no-na, or in proper Finnish, Vi-no-na. It still reflects back to its glorious days between 1898 when the copper mining company was employing more than 300 persons, and the closing of the copper mines on March 15, 1920, the period during which the community saw active copper mining operations.

Winona's name is of Indian derivation and is referred to in some of Longfellow's works. Winona's discovery was via many Indian pits which were found in a specific pattern. They revealed that much copper was taken out, with no indication as to its possible use. Years after the site was discovered, hundreds of copper finds were located in the Winona area, all beginning in the nature of outcrop severances. Finds of float copper in the area have been rare.

The Winona Copper Company was made up of many immigrants who came to settle in the Copper Country. These consisted of miners and foremen from Cornwall, England, and Finnish, French, Croatians, and Italians, according to the Company's payroll.

Employee housing was completed in 1864 when the first copper mining company built a large boarding house of hewn timber with matching floors and a shingle roof, and one small dwelling house, built the same way. When the

HERE is a mine site near Winona. Obviously, the structures no longer are in existence but the buildngs do reveal how extensive the operations were during their period of activity, 1900 to about 1920.

new Winona Copper Company started in 1898, they used the existing facilities and did not build any employee housing until 1903. That year, two double dwellings were erected at the cost of $2,470.00, giving accommodations for four more families.

According to an existing map, six major locations comprised the village; Log or Finn Town, Central, King Philip, Shoe Pac Street, Wall Street, and Red Location.

The year of 1906 found much construction in the village of Winona and the King Philip Location. A new school was being built by C. C. Davis, and a number of other buildings were being erected by Contractor Davis. An order had also been received by the Maass Brothers to build nine houses for use of employees of the Winona Mining Company. These wood-frame structures were substantially built, and better than the average home for a mine worker.

The Winona also had three boarding houses erected, 36 by 24 feet, 26 by 28 feet and 24 by 34 feet respectively, the latter with a basement and stone foundations. The two others had post foundations. Three boarding houses cost the Winona $3,676.18.

One of the three structures mentioned above, burned to the ground on Saturday, April 15, 1916, when fire broke out in the boarding house maintained by Frank Daniels at 2 o'clock in the morning. The structure

ELM RIVER TWP.

WINONA

SECTION 29, 30 T52 R 36

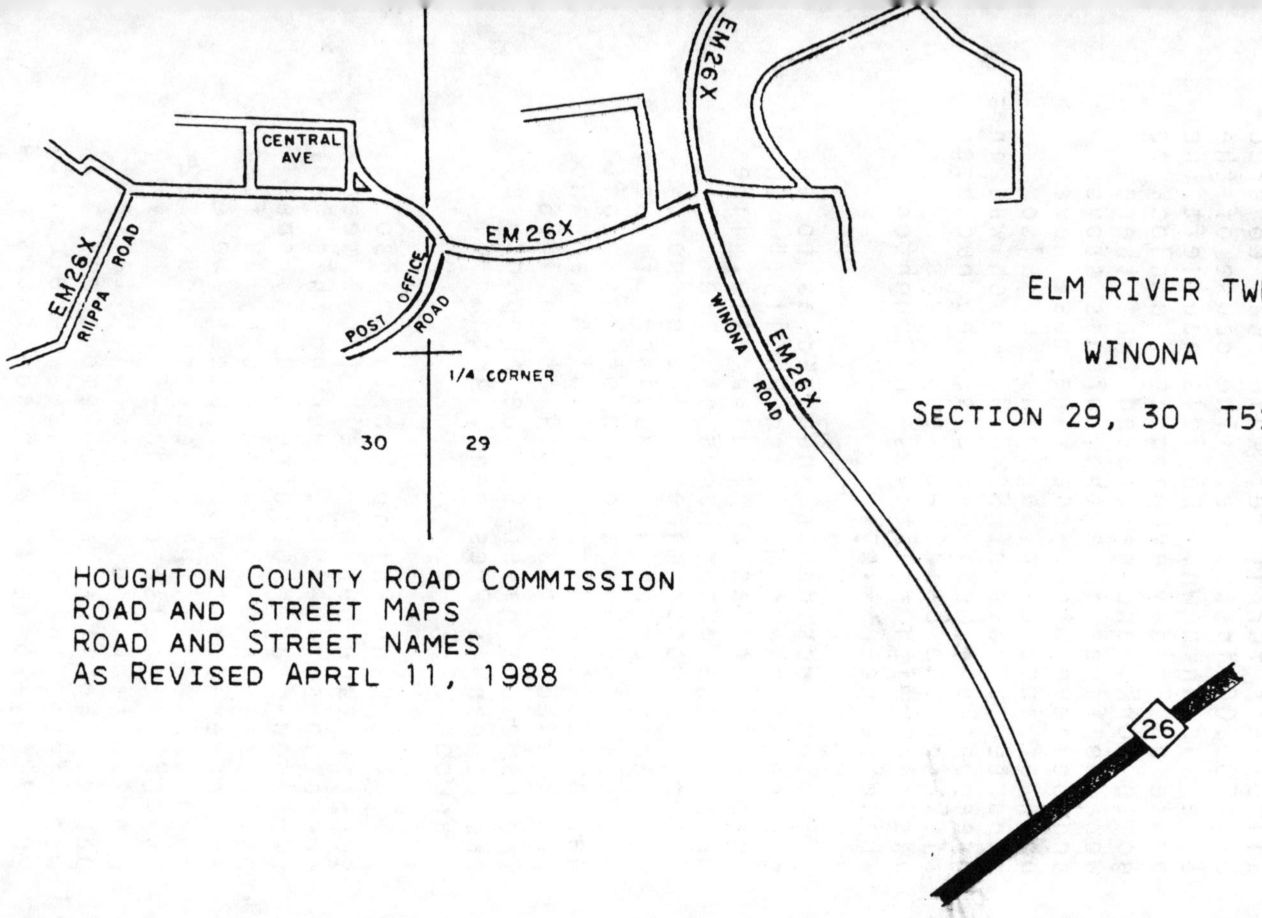

HOUGHTON COUNTY ROAD COMMISSION
ROAD AND STREET MAPS
ROAD AND STREET NAMES
AS REVISED APRIL 11, 1988

was owned by the Winona Copper Company and was used as a boarding house for its Italian employees. Daniels lost all of his furniture, which was covered by $600.00 insurance. The cause of the fire was unknown. Daniels had left the house on Friday evening at 7 o'clock to go to work, and he stated that there were no fires in either of his stoves and believes that the fire must have been of incendiary origin. He also reported a loss of $180.00 which was in greenbacks. Mrs. Daniels, who had been sick for the past two weeks, was visiting with relatives in Houghton while recuperating.

Some residents inquired as to the source of the name "King Philip." They were told it was an Indian name, like many of the other names of Lake Superior copper mines. The property was named for Philip, advisor of the Wamanoag Indians and son of Massasoit. He caused the "King Philip's" war and was killed in that conflict in 1676. The name was given to the property by the Company's President, Arthur G. Stanwood.

The King Philip Location also saw construction during their first year of operations, in 1906. During this year two wood-frame boarding houses, one 15 by 18 feet and the other 20 by 40 feet, and covered with tarred paper, were built, along with a sleeping house 22 by 40 feet, also of wood-frame construction and covered with tarred paper. For the married employees, 12 wood-frame dwelling houses were built, 20 1/2 by 26 feet, with additions of 4

1/2 by 10 feet, with stone foundations and cellars. These 12 homes were built in block 1 and 2, between the pumphouse and shaft No. 1. The Captain's office, men's camp, boarding house, dry and power house were nearby.

Among the other buildings constructed at the Winona during 1906 were carpenter, blacksmith and machine shops, and a new residence for Superintendent Rex R. Seeber, which was ready for occupancy in early fall. This structure contained eight rooms, was modern in every respect and cost approximately $4,000.

It was also rumored that a number of miners who were at one time employed at the Atlantic Mine were comming to Winona during the summer months to seek employment. The new store manager would also renovate the store that he had recently purchased and make it much larger.

The Winona Mining Company's crew of carpenters were also at work constructing a new barn during January of 1906. Seems that their barn burned to the ground on Monday evening, December 25, 1905. Eleven horses were burned to death in the fire which broke out in the barn of the Winona Mining Company at Winona location. The fire was first discovered around midnight, and it had already been in progress for some time. It was impossible to do anything to save the building or its contents which consisted, besides the horses, of several tons of hay. The eleven horses were all fine animals and their loss was a serious one. It was serious one. It was estimated that the total loss including the barn and horses, was about four-thousand dollars and was partly covered by insurance.

As mentioned above, the new barn was built by the company's carpenters. Labor to erect the new barn was $108.00; the remainder was covered by insurance. Horses were purchased from Mr. Saxby of Hancock and operations were being carried on as in the past. The barn was completed in two weeks, according to the _Evening Journal_ of Hancock.

During 1907 the Winona built 20 six-room single family houses, 2 stories high, each 20 by 26 feet, with 4 by 10 foot extensions on the rear which were used as a pantry and a vestibule. Two seven-room single family houses, each 24 by 30 feet, with a 4 by 12 foot extension on the rear for a pantry and vestibule. They all had stone foundations, basements, running water and sewer. According to the map, they were built on blocks 4, 5, 6, and 7, near the school house and dispensary, but between the No. 3 and No. 4 shafts.

According to the annual report for the King Philip, more homes were built in 1907; however, only 13 homes are shown on the map. The homes in the King Philip location were furnished with electric lighting at this time, but not so at the Winona.

The year 1909 found three homes built for the Winona mining officials.

THE King Phillip location in the Winona area. Copper Country people today seem not to realize what populations were contributed by the smaller mining communities of Houghton County. Once with a population of 88,000, Houghton County relied on such areas as the King Phillip to swell its overall populational worth. Today, there is nothing like this avenue in the Winona region.

A ten-room dwelling, 28 by 34 feet, was constructed for the mining captain, and two ten-room frame dwellings, 26 by 38 feet, for the mine physician and stamp-mill superintendent, respectively. Each home was supplied with a bathroom, hot-water heating plant, electric lighting and a company telephone. These homes were built behind the school, on the street locally known as "Bosses Row."

The Winona Mining Company, like other local copper mining companies, had a physician who treated their workers and their families. The Winona dispensary was built during the spring of 1906. It was a wood-frame building, 18 by 32 feet, two stories high, with a stone foundation, but no basement. There was room for three or four patient beds on the second floor. This structure cost $722.40 to build and was located between the Nos. 3 and 4 shafts, just south of the location and and behind the new mine office.

Dr. A. D. Aldrich moved to Winona from the Challenge Mine during May of 1906. He had held the position of physician there for several months, and before that had been the assistant mine physican at Painesdale. With his main office in Winona, Dr. Aldrich was now in charge of the Challenge, King Philip and Winona mining locations.

Dr. Aldrich while a young man was a very able physician, and the company recognized the efficient work which he was doing at the Challenge and thought him worthy of the promotion. Dr. Aldrich had an assistant a short time later. He started his new duties on May 1, 1906. He was followed by Dr. Cox and Dr. Felton.

A ten-room wood-frame home, two stories high, 26 by 28 feet was built for the mine physician during 1909. It had a bathroom, a hot-water heating plant, electric lighting and telephone service. As with other Company houses, it had a mine-rock foundation and a full basement. His home was one of the three built behind the school in "Bosses Row."

The annual report for 1910 showed that the Winona added a fire department to the location at a cost of $1,300.00, for the building and equipment. This wood-frame building was supplied with electric power and water. The fire department was located near the dispensary and office, and the building was also used as a morgue. The men's bodies were brought here after a fatal accident to await transfer to their homes. It was customary to bring the body to the decease's home for viewing by relatives and friends for one or two days, before being brought to the church and then burial.

It was an unwritten rule at all copper mining locations that at the sound of the fire bell, all men would respond to the alarm and go to the fire. Only those men who could not be released from their surface jobs or who were in the mine, were excused from answering this emergency alarm.

The employee's location obtained drinking water from the No. 1 King Philip shaft. The water came from an underground room-like portion of the mine, from a fresh spring water source. A pump station, 22 by 18 feet by 11 feet had been cut 135 feet below the collar of the shaft, and during 1911 an electric pump was installed. This project cost the Company $1,499.42. The water tank was built on top of the hill, near "Red Location", where the steel tower is now situated.

The one reason for the Winona name to persist is the still-operating school in that community. There is little doubt that if the Elementary school were to cease operation in the old-time mining community, the traditions behind the town also would go and tend to be forgotten. Today, Winona is the most westerly operating school in Houghton County.

It was on May 4, 1906, that the residents approved a new one-story, wood-frame schoolhouse at Winona. It was approved by a unanimous vote of all taxpayers of Elm River Township, at a special meeting held at the office of the Wyandot Mining Company. A three-mill tax was ordered levied on all taxable property in Elm River Township for the erection of the three department schoolhouse at Winona.

June 29, 1981

The Winona School as seen on June 29, 1981

According to a newspaper article published in the _Daily Mining Gazette_ on May 24, 1906, C. C. Davis, who had secured the contract for the erection of the new schoolhouse, left Painesdale the day before for Winona, to start work on the new structure. Mr. Davis expected that the work of erecting the new building would require about three months. A large force of men was set to work at once so that the new structure would be ready for occupancy at the beginning of the new school year.

The large attendance at that school made it necessary that the school should be ready by the first of September, and Mr. Davis was certain that he could comply with their wishes and would do all in his power to see that the work was finished.

The old school was a large log house that the Township had been renting from the Winona Mining Company. The building was now too small and was very hard to heat. The old log building was to be used for other purposes, and should the student attendance so increase that a larger building be necessary, the old building could be moved and joined to the new one according to Township officials. The teachers for the new school had not been engaged for the coming year, but that would be done the latter part of May and the first week of June 1906. The teachers moved in at the end of

August of 1906, and school was begun as scheduled in September. This new building not only replaced the old Winona school, but also the small school structures in the farm regions of Donken and Stackpole.

Due to the growth of the town, two stories were added to provide the school district with a school to house all twelve grades. This included a gym on the second floor. It was in 1910 that the older Elm River Township students were able to attend high school classes in Winona.

A commencement announcement dated June 14, 1916, shows that the class motto was "Prepared for better things," their class colors were dark green and white, and the class flower was an American beauty rose. High School graduates were Thomas H. Thomas, Lucretia Maine Reichert and Ida Sophia Perquist. The school orchestra played four selections during the program.

The Winona high school students had a very active baseball team during 1930, as they beat the would-be champions at Rockland. The Winona students also visited the Copper Country Fair on September 25th, 1930, which was Children's Day.

Church services were conducted in the old school, and then moved to the new school when it was completed at the end of August, 1906. They later held their services in the nearby auxilliary building after it was completed.

Catholics received permission from the School Board to conduct their services in the High School, and even had a choir directed by Mrs. Fournier.

A second wood-frame two-story building was constructed during WPA days, alongside the old school. It was built by the school district with government aid. Former Superintendent Emery Jacques worked diligently to obtain this school for Winona. He later became warden of the Marquette Branch Prison.

The first floor contained classrooms while on the seond floor there was a gynmasium and an apartment that used to come with the job of Elm River School District Superintendent. The gym in this facility included locker rooms and showers. It had a much larger gym than was in the old school.

The gym floor was built during the fall of 1934. It was first tested when the Winona High School basketball team met the Sacred Heart High School of Calumet. It was the first game scheduled on the new gymnasium floor and played on Friday, November 16, 1934. The gym was high school regulation in size, 66 feet long and 40 feet wide. A bleacher section and balcony ran the entire length of the court. Mr. Schon of Ontonagon refereed the game that night.

This school had a champion basketball team, and the members traveled to other local schools to play, usually traveling by a Copper Range Railroad

September 20, 1990

Inside the Winona School

special. The building was also used for social activities, and elections were held here. During the building's final days it was used as a community building. This auxiliary building fell into disuse and was salvaged in 1980.

During the 1979-80 school year, the Elm River school had 26 students and three of them graduated from the 8th grade with appropriate ceremonies. The graduates were Stuart Maki, Patricia Leppanen and Shelly Gourneau. Principal Steve Ipple gave the commencement address before towns-people from Winona, Donken, Twin Lakes, Bohemia Township and Toivola. The graduates then proceeded to the Jeffers High School in Painesdale to complete their high school work.

Winona's school, on the last Friday of September 1980, had 25 students in grades primary to late elementary, the latter being the eighth grade. During 1980, the principal was Ashley Luplow who resided in Alston. The primary teacher was Christine Lickman and aides were Mrs. Eva Lindeman and Flora Knudsen.

Several newspaper articles announced student finalists in Q-Tip art under the instruction of Flora Knudsen. In June of 1983, four students were named finalists in a national art contest that drew 40,000 entries each year. Winners were kindergartner Avery Paavola who won a $500 savings bond, with the others

Miss Carlson's 4th, 5th and 6th grade class, taken on Monday, May 16, 1927 - courtesy of the Harold H. Heikkinen collection, Atlantic Mine, Michigan

being kindergartner Alan Huhta, sixth-graders Greg Leppanen and Todd Auel. Their works were displayed at the Metropolitan Museum of Art in New York City.

The next year, 1984, winners were Casie Huhta, Paul Mickalich, Sheila Snell, Ryan Gourneau, Avery Paavola, Alan Huhta, Danielle Gourneau, Brandie Snell and Jon Paul Maki. Their art teacher was Flora Knudsen. Two of the paintings one of the mine and another of the school, still hang in the school.

Today, the main school building is one of the few wooden structures that has not fallen victim to time and disuse. It still looks much as it did when it was built in the 1900's. At this time only the basement and first floor of the three-story building are still in use.

It is no longer topped by the school bell that once announced the start of classes. According to a former student, they used to ring the bell four times a day, at quarter to nine to let the pupils know school was starting in 15 minutes, at nine, at quarter to one and at one o'clock. The bell has been removed and given to the First Lutheran Church for use at that location.

The school district no longer offers a high school education. Students attend the Winona school

July 9, 1978

This auxiliary school was built during the WPA days

July 9, 1979

through eighth grade and then transfer to Jeffers High School in Painesdale. During the 1992-1993 school year, 24 pupils attend this school. The two teachers are Christine Lickman and Paul Halonen. Jeanne Gourneau is the teacher's aid and secretary, Debbie Short the cook, while David Usitalo is the custodian and bus driver. Once a week an art teacher and a musical instructor also teach here. Once a month, the district's school bus takes all the children to the Portage Lake Public Library where the students can check out books or do some class research.

The Winona First Lutheran Church was built in 1958 and the edifice is located immediately across M-26 from the entrance to the old Winona and King Philip mining firm's town. The white frame building is easily observed by the motorist proceeding between Painesdale and Mass-Greenland. It is some 1,000 feet off the highway and is not far from the Winona Cemetery which is reached along the same road that serves the church.

The edifice is said to have cost approximately $25,000, although much of the labor and material was donated. The indebtedness was cleared during August of 1963 through the sale of pasties to persons in the Twin Lakes - Winona regions. The church board has a list of the benefactors. In about 1963 an Elm River Township resident contributed an electric organ. However, the current organ was donated

September 12, 1992

First
Lutheran
Church
of Winona
Sunday
11:30am
September 12, 1992

by the parishioners of a church in Iron Mountain. The present church organist is Flora M. Knudsen, who also assists other churches in the area who request her services.

The church is a homey, simple building which capitalizes on plainness for its utility and attraction. Seating 100 worshippers, it also has a convenient basement dining room and kitchen which makes it possible to stage socials and dinner parties. The upstairs or church auditorium also features two sacristy-type rooms. One of these is the clergyman's vestment room while the other is used for storage of church equipment. The two tall stained glass windows behind the altar were designed by Flora M. Knudsen, and they were installed in the memory of Mamie Poyhonen. The small eight-sided window behind the organ, with the blue butterfly, is in memory of Bart Autio.

The new entry was just built in 1990 and was designed by Albert Leppanen. He included a handicap ramp into the main auditorium, eliminated the outside steps, and changed the entrance to the basement kitchen and dining rooms. The bell that was removed from the school several years ago is now located in the basement furnace room waiting for a tower to be built alongside the church to house this antique.

The building may be entered from the side or rear. The auditorium area

September 12, 1992
First Lutheran Church of Winona

September 12, 1992

is nicely furnished with pews and all necessary equipment required for regular church services. During the summer, the church features services every Sunday. During winter they are less frequent.

As mentioned above, much of the building was paid for through pasty sales, about 200 being made each week. Mrs. Ida Saari headed the pasy-making group, assisted by Mrs. Floyd Lindeman, Mrs. Elvin Lindeman, and Mrs. Frank Wiita. John Peterson was in charge of disposing of the finished pasties.

The parish is now being served by Tom and Sharon Huff, who succeeded Thomas Macey. Donald Reichers served before him.

The Winona cemetery is located in section 32, just off Highway M-26. Although residents are not sure just when the cemetery was started, an old map of the Woodland Cemetery is dated 1914 and the cemetery was begun a few years before that date, in about 1910. It is operated by Elm River Township, and at this time Township Supervisor Art Dasse is the custodian, opens and closes the graves and completes the necessary paperwork. Elmer Katajamaki owns the property that the cemetery occupies, part of his 80 acre parcel.

It is said that the first tombstone was donated by the community, for a child that was one year old, born in 1910 and died in 1911. It is the tallest monument in the cemetery, about

September 12, 1992

September 12, 1992

four feet high, made of gray granite. Engraving is in the Finnish language, has a cross on the top, with the name "EVONSON" on the bottom. The grounds are kept neat, grass mowed, and the leaves and sticks picked up.

The Finnish Temperance Hall was built in 1906 on Company-leased property and located in "Red Location", between Camp No. 1 and 2, block 9, across the street from "Finn Town." Finn Town conisisted of 8 or 9 single family log homes, all in a row, which were usually occupied by Finnish families. The Finnish Hall was a two-story wood-frame structure, built and owned by the Society, where they conducted their social affairs such as dinners, shows, lectures, meetings and various gettogethers. Finnish Lutheran services and Sunday School were also held in the Finnish Hall.

The men also had dinners, one by the community's bachelors and the next time by the married men, and so forth. It was a community affair and everyone was invited to attend. They also had a gym with weights, trapeze bars, and other types of physical equipment so the men could work out. The location's Athletic Club held their weekly calisthenics here. The Company was very supportive of the Finnish Temperance Hall, as it gave the working man something else to do besides drink at the local saloon.

National Archives' Record of Appointment of Postmasters shows that Winona's postal service was established on April 5, 1899. It was about the same time that there was a revitalization of the old Winona Mine and the Copper Range Copper Company, whose railroad the company was using to transfer copper ore. The postoffice was temporarly discontinued on June 20, 1921, but was reestablished on July 8, 1925.

Postmasters serving the citizens of Winona are as follows: George McElveen, 1899; Joseph Paul, 1899; Thomas J. Kearney, 1904; Jesse Daley, 1906; Daniel Wessels, 1908; John McCarthy, 1919, Anthony Grinditti, 1919; Theodore Wiedenhoefer, 1919; Pascal Procissi, 1920; Nell Rowe, 1920; and Frank Kline, 1925. Mrs. Sophie F. Kline was commissioned on June 5, 1931, and assumed charge on June 30, 1931. Francis Gauthier was was appointed acting postmaster on August 3, 1944, and assumed charge on May 19, 1944. Mrs. Taimi Gauthier was the last postmaster who assumed charge on January 31, 1945, and was commissioned on September 11, 1945.

This Fourth Class post office was closed on June 20, 1920, and the mail was then sent to Donken, and when this post office was closed, mail was sent out from Toivola.

The old post office building burned down 57 years after it was closed, on Saturday, August 27, 1977. The building was located behind the

Winona Elementary School. Fire gutted the two-story structure and only the lower story of the building, which was once the official home of the Winona Copper Company, remained. The fire was believed to have started from lightning. It was just after a loud crash of thunder that the fire was spotted by Mr. and Mrs. Willard Cossey of Greenland. They informed the Wilbert Leppanens of Winona, who notified the Twin Lakes Fire Department. The blaze began in the upper part of the building.

Some furniture was removed before the blaze became intense. Much of the equipment was consumed in the fire. There was no estimate as to the amount of insurance carried.

Three departments fought the flames, Winona-Twin Lakes, Laird and Greenland. The fire was under control about 5 p.m. but the Winona-Twin Lakes unit came back at 11 p.m. to inspect the premises.

This building had been put into use as a postal facility immediately after the Copper Range depot site was abandoned. The last postmistress and her husband owned the building.

One of the wood-frame homes on the main road, with a square front, was a barber shop. The front addition is rather recent. This barber shop was used during the mining days, when miners got their haircuts in this

This home was once the location's barber shop

structure and then went across the street to the saloon. It was owned and operated by a Negro by the name of A. H. Martin. His shop was in the front room, with his kitchen, bedroom and bathroom in the rear. The shop, with high ceilings, has hardwood floors, and circle marks from the barber's chair can still be found on the floor.

Mr. Martin was mentioned in The Keweenaw Miner when a Winona resident, Charles Waisanen, broke into his shed and stole a pair of trousers belonging to Mr. Martin. Waisanen had to pay the costs of his prosecution and reimburse the barber for his trousers. The home is now owned by Uno Sarja.

The saloon was located across the road from the barber shop in a two-story wood frame building. Owned by the Bosch Brewing Company of Lake Linden, the saloon had a large bar, a mirror which ran along the front of the room, with lots of spittoons arranged around the room. Of course, they only served Bosch beer. The front half of the first floor was used for saloon business, with the back half used by proprietor Paul Fournier and his family. The second floor had hotel rooms which were used by traveling salesmen, visitors to the town and newly-arrived workers who were looking for a place to live. Mrs. Fournier made and served the meals that were included in the charges and was in charge of cleaning the rooms. The large wood-frame building had the usual square front as other businesses in the village. The Bosch Brewing Company also had a warehouse near the railroad depot. All beer was delivered in wooden barrels. Charles Kratt was the distributor.

This building was sold in 1922 and turned into a grocery store. It was owned by Kalab Seppanen and operated by his daughter Betty. Here the customer could purchase general types of groceries and kitchen items. After catering to the village's wants for many years, Mrs. Seppanen finally closed the store in 1961. It was true that declining business was a factor, but it was surprising how much business came from tourists, hunters, fishermen and visitors to the community.

All mining companies built and maintained general stores for their employees. Here workers and their families could purchase groceries, meats, bakery, household supplies and even clothing and furniture. Most of the purchases were charged and the amount was taken from the employee's wages before he received his pay envelope. Winona was no exception, and their store was built and located in "Red Location," on the top of the hill, which is now a large clearing. The reason this area was called the "Red Location" was that all of the homes were painted red.

The first company store was built in 1899, but burned in February of 1900. Later a new store was rebuilt by Company carpenters, along with a small ice house and a meat market. A few years later, in 1903, the store was painted inside and out, amd a cellar was constructed under it with stone walls and a concrete floor.

After several years, the Company sold the store to Mr. H. S. Goodell who was secretary of the South Range mercantile company and resided at South Range. He closed the deal for the general store on May 19, 1906. The store had been under the management of Mr. Carney for the mine owners. He had operated the store in the past and continued in that capacity. Mr. Goodell had resided in Painesdale for a long period and wanted to continue his residence there.

Winona residents were told that their store would have no connection with those owned by the South Range mercantile company at Baltic and Painesdale and would be owned solely by Mr. Goodell. Conditions pointed to a very successful summer at the Winona mine and the new acquisition would prove to be a successful purchase for Mr. Goodell.

The new store at Winona would be conducted upon the lines of a general store, and would operate on the same basis as the stores with which Mr. Goodell was already connected. It was not known at the time of the purchase if Mr. Gooodell would change the structure, as the building was in good condition. It was by no means an old structure.

Next to the school and the church, Jake's Bar is one of Winona's best known businesses; in fact, the only business. Located on highway M-26, just at the turn-off to Winona, it was built by Charlotte and Jake Erickson in 1948. They operated the bar for 34 years. Besides liquid refreshments, they also served snacks and hamburgers. The Ericksons lived in a home next to the bar.

It was operated on a lease basis for a time by Ross Kischer, who moved to Ontonagon and now operates Doc's Bar.

The current owner is Roy Bowling who purchased the property from the Ericksons on September 1, 1983. He obtained the bar after traveling from Canada one summer, passed this way and saw the bar. He remodeled the interior and made it a sunny comfortable place that invites relaxation.

It was described by a Mich Tech University LODE reporter "--- as a roadside bar, and as a house that has had the downstairs renovated with a bar and pool table installed, Jake's is a very simple place. It's got your basic bar, booze mixed with pop or beer. There seemed to be a small number of clientele but obviously very regular and it seemed like a friendly place." He ended his article by saying "Jake's is definitely a good place to stop on the way home from skiing at the Porkies."

When entering the well maintained and clean establishment, the customer finds the main bar on the left hand side of the room, running its entire length. In the middle of the room are two side wooden conversation bars with stools. The patron can obtain hamburgers, pizzas and chili from the small kitchen. A juke box, regulation pool table and cherry-wood piano also make up the room. Roy lives in a two bedroom apartment to the side of the bar, also on the main floor.

The well-known cougar still resides in his cage at the rear of the building. "Thor" weighs in at a little over 235 pounds, heavier than the average wild cougar. He eats six cans of dog food and seven to eight pounds of meat each day. During deer and bear season he eats his fill of scraps brought in by local hunters. Today he stays in a specially designed and insulated cage out back of the bar, amusing himself with a tire swing, balls, and stuffed animals. The cat's extra-heavy winter coat and lots of hay in his house keep him warm all winter long. In the summer the hay is removed so Thor can stretch out on the cool cement floor.

Just in front of the school one finds a World War II Honor Roll. The memorial is four feet seven inchs wide, by six feet eight inches high. The Elm River Township white wood-built honor roll sits on a cement base. It

September 12, 1992

contains 45 names, with three men listed as killed in action. The three are William Lehto, Peter Sterbis and John Mc Laws. The honor roll is scheduled to be repaired during the summer of 1993. Flora Knudsen will be doing the printing and art work.

All mining communities stressed sports, as these were the resident's main source of entertainment. Some forms of entertainment in the Winona region were: dinners, baseball, dancing, and at one time they had the rudiments of a good band. The school usually had an orchestra.

They also had one of the best baseball teams, with the late John McCarthy of Houghton, a former clerk at the Winona Mine, being their catcher. Even during the later years of his life, he played behind the plate and always sang the praises of the town at whose mining firm he enacted a clerk's roll.

The Winona ball field was located behind the No. 3 shaft, was surrounded by a wooden fence, and had a grand-stand. There was a place for the baseball players under the stands. The property, fence, stands, and equipment were furnished by the mining company.

Several residents still recall Winona's baseball team, as they watched the men play and occasionally followed the team as it moved hither and yon via the Copper Range Railroad.

September 20, 1990

One of the items listed under "Winona News" in the May 6, 1916 issue of The Keweenaw Miner stated that on Tuesday evening several of the Winona baseball fans got together at the mine office and began to make arrangements for a baseball team which they expected would be better than the one they had in 1915. They were going to hold a public dance in the near future to raise money and start the team out with a financial standing. It gave them initiative and showed them that the residents were behind them heart and soul as they were in past years.

Prominent fans had endorsed the stand taken by James Gregory of Atlantic, who suggested the formation of a league and criticised the fans for not supporting baseball as they should. Many people asked why should Gregory desire the formation of a Range league and let Winona out? Some fans would "holler" because they could not make a little money out of their ball playing and some fans would state that expenses were too high if they were compelled to travel out of town for a game.

Winona was short of fans in number, but had them in quality, a good example being the erection of the local baseball park, the only one south of Hancock. The article stated that they were to give another hurrah for Gregory, but whisper to him the following: "Organize your league, but

TWO weeks ago the Green Sheet used a photo of the Winona baseball team. Larger than this one it featured a later squad that represented Winona on the baseball diamonds of the Copper Country. This one, with John McCarthy, late justice of the peace, at the head, is one which represented the Elm River Township nine in earlier years. It is from Mrs. Otto's memorable files.

consider your neighbors, tho they be but a few miles away."

Winona was provided early railroad service by the Copper Range Railroad Company. During 1902, a spur of the Copper Range Railroad was built to the No. 2 shaft, the only shaft then in use, with the spur being 3,600 feet long. The roadbed was built by the Winona Copper Company at a cost of $1,030.68. Labor was $919.48, explosives $53.50 and tools and supplies $57.70. The track-laying and surface work was done by the Copper Range Railroad Company at their own expense.

A new Winona railroad spur 3,755 feet long was built during 1906, from a point on the old spur to and beyond the No. 3 shaft at a cost of $2,533.91. It connected with the coal trestle and coal bins. The Copper Range Railroad also received $3,416.36 for transporting rock to the stamp mill.

The next year, 1907, 2,334 feet of Copper Range Railroad steel rail was extended to the No. 4 Winona shaft and new shops at a cost of $5,040.91.

Before the King Philip Mining Company became part of the Winona, they were also serviced by the Copper Range Railroad. When the Company was starting out in 1907, construction of 3,336 feet of sub-grade was completed for extension of Copper Range Railroad tracks from Winona past the No. 1 shaft to the Compressor plant.

DEFINITELY nostalgic is this picture of the former Winona Copper Range Railroad station. Many is the person that traveled to Houghton, Ontonagon and points as far south as Milwaukee and Chicago via a wait at this station. Now, just as is the rail line, it is no more. Tne near disappearance of Winona serves to reveal how the Copper Country has lost its population, via the near elimination of the small community such as Arcadian, Lake Mine, Phoenix, Jacobsville, Redridge and Rhode Island.

During 1911, a building was erected on the Winona property, near the machine shop to house the Copper Range Railroad steam locomotive (number 102) and the electric locomotive (Ojeebik). The electric locomotive looked like a trolley car, but was reinforced to do the rough railroad work. The round house cost $1,563.04 to build.

As the years went by, spurs were extended to various parts of the Winona and King Philip properties, and finally extended to the compressor plant just past Sleepy River, to the far southwest. The King Philip No. 2 shaft was located here, near the Houghton and Ontonagon County line.

The Copper Range Railroad Company built a number 1 wood-frame depot at Winona in July and August of 1907. A three page bill of mill work, dated July 26, 1907, was recently found in the Copper Range Railroad Company archive file at the Michigan Technological University Library. It was a large wood sided building with the center section being two stories high. The station manager and his family lived on the second floor. The structure provided both passenger and freight services. The depot was located at the southeast section of the location on the main line, across from the employee's Red Location.

Winona residents used the Copper Range passenger trains regularly and it was surprising how active the station was during the days when the mine was running. Most often the trips were to Houghton and Calumet, but there were many times the patrons moved to Ontonagon and Mc Keever for transfer to Milwaukee and Chicago. It must be remembered that almost everything came to Winona by train, and box cars were spotted at the depot to await the transfer of merchandise and freight to the stores, warehouses and the various mine buildings.

WINONA COPPER COMPANY,
F. W. DENTON, Agent.
WM. C. VAN ORDEN, Clerk.

WINONA, HOUGHTON CO., MICH. Jan 13/06

The original Winona Mining Company was organized in April of 1864, and during 1864 was fully engaged in active explorations and improving the surface of its estate. The estate consisted of six hundred and forty acres of land in Sections 22, 29 and 32, T.52 of R. 36 W. The credit for the discovery of this property went to Joel Tresider, the man who first found and opened the "Tresider Lode" at the Bohemian. He was justly proud, and well he may should, at this second instance of his success in exploring.

These lands were situated some two and a half miles east of the Penn mine, and were formerly owned and recently organized by Messrs. G. C. Jones and

Jay A. Hubbell. Among the results accomplished that year. Were the opening of promising cupriferous deposits, the erection of several dwellings and other buildings.

The openings were marked by a continued series of ancient mining pits along the outcrop, invariably regarded as among the most favorable indices of the value of a lode. After examination of these old workings in all parts of the mineral region, they came to the conclusion that they had been made at different periods, by two different races of people, one of whom simply removed the soil, and with the help of fire and their rude stone hammers, pounded out the small horns of copper that protruded from the outcrop of the deposit. The others had sunk holes several feet into the rock, had driven in short levels, or rather burrowed into the rock a considerable distance.

These ancient pits were cleaned out in many places, the back of the lode stripped, a shaft started (which was down about fifteen feet in the vein.) The copper mined during these operations had been more than sufficient to pay for all of the mining expenses that year. Besides heavy barrel work and coarse stamp copper, some pieces were of thirty pounds weight, and nearly pure mineral. Fine specimens of native silver were being taken from the openings on this vein, and a large proportion of the removed rock was thickly charged with silver. Captain Grierson was in charge of the mine.

The Portage Lake Mining Gazette reported on October 8, 1864, "This vein of such splended surface exhibit, opened several months since, and from which nearly two barrels of copper has been taken, and no shafts yet sunk! The vein looks well where the workmen were engaged clearing out for a shaft."

This same newspaper published on November 26, 1864, "--- have stoped out at one point where there were some ancient workings, the full width, and about 25 feet in length, and about 300 tons of rich stamp and barrel copper taken out. The shaft is now going down as fast as possible."

The surface improvements completed during 1864 consisted of building one large boarding house (of hewn timber and with matched floors and shingle roof,) one small dwelling house, one blacksmith shop, one stable, one warehouse, timber, lumber and a coal pit. In addition there were sufficient supplies on the ground to last until January 1st, 1865, when transportation would be more cheaply and readily available. The Company had eight thousand dollars in the treasury, and copper already mined valued at $3,000.00 or more.

The Winona property was heavily timbered and had water privileges which were important to the development of the mine. A branch of the Sleeping River crossed the south-west corner of the west quarter, affording excellent facilities for stamping purposes. Jay A. Hubbell signed the annual report for the Proprietors.

The Lake Superior Miner wrote on January 7, 1865, "This time, the barrel and stamp work both seemed more profuse and richer, if possible, than heretofor. Chunks of vein rock of 5 to 50 pounds weight, with 10 to 50 percent of mineral, are abundant."

On April 28, 1866, The Lake Superior Miner published "During the latter part of the summer and fall, several very favorable proposals were made to the principal owners of the property to take it on tribute, one of which was accepted. Two men, inexperienced in mining, went there and commenced work about January 10, 1866. No shaft house over them, they worked only when the weather was reasonably good, and they cleaned up in last month, what is estimated at 4 tons of 75 percent mineral. This was barrel work taken from a shaft 25 feet deep by 15 feet long and 12 feet wide. There are now two shafts partially sunk on this vein, about 600 feet apart, both of which have yielded remarkably well." Owing to the lack of transportation facilities little was accomplished and the mine was soon closed.

INCORPORATED UNDER THE LAWS OF THE STATE OF MICHIGAN.

Countersigned. American Trust Company, Boston. TRANSFER AGENT.
by [signature] Assistant Secretary.

Boston Safe Deposit & Trust Co. REGISTRAR OF TRANSFERS.
by [signature] Assistant Secretary.

NUMBER A5693

SHARES 100

Winona Copper Company

CAPITAL STOCK $5,000,000.
IN SHARES OF $25 EACH.
$25 PER SHARE PAID IN.

This Certifies that CHARLES E. PERKINS, HOWARD ELLIOTT, NATHANIEL H. STONE TRUSTEES UNDER WILL C. E. PERKINS is entitled to ONE HUNDRED Shares of the Capital Stock of the WINONA COPPER COMPANY transferable only on the books of the Company by the holder hereof in person or by attorney upon surrender of this certificate.

Boston, NOV [illegible] 1918

[signature] Secretary. [signature] President.

From The Portage Lake Mining Gazette, dated June 10, 1880, "It is reported that some mine work will be done the coming season on property known as the Winona."

This same newspaper printed on August 25, 1881, "At the Winona mine the two old shafts have been cleaned out, and the eastern one sunk to about 20 feet deep, where the vein is found to be richly impregnated with copper occurring in the shape of small masses, as well as in the rock suitable for milling. The shallow pit has not yet reached the settled portion of the vein and of course does not make as good a showing as the other." The men holding the lease and and working the mass copper soon found it did not pay and closed the mine.

The Ontonagon Herald printed on Aguust 19, 1881, that the work was principally of an exploring nature, and the force was small, but the results thus far were highly satisfactory. It was probable that the owners would put the mine on the list of live industries or place it in the hands of capitalists for this purpose. Mr. G. C. Jones, Jr. had charge of the mining works.

A year later, September 21, 1882, The Portage Lake Mining Gazette published "A thorough system of exploratory work will be inaugurated soon on what is known as the Winona mineral estate, situated in the southwest end of this count in T 52, R

36. The prospecting is to be done by parties who intend to purchase and mine the property."

The first annual report was published in 1899, 35 years after the Prospectus of the Winona Mining Company of Michigan was published. John Stanton was President of the Company, John S. Stanton as Treasurer and J. W. Hardley as Secretary. Directors listed were John Stanton, Joseph E. Gay, William A. Paine, J. R. Stanton and James H. Seager. They had one hundred thousand shares of stock, sold at $25.00 per share.

This Board of Directors also controlled the Wyandotte Copper Mining Company which joined with the Winona on its east, and the Escanaba Mining Company which adjoined the Wyandotte. These three properties totaled over 1,400 acres of land, having a length on the vein of over 1 1/4 miles according to the <u>Ontonagon Herald</u>.

The annual report covered the months from November 1, 1898, the date that the Winona Copper Company took possession, to December 1899. At this time the No. 1 shaft was 97 feet deep. The No. 2 shaft, just south of the No. 1, was sunk 45 feet and covered with a small shaft house and operated by a small hoisting plant. No large masses had been found and but very little barrel work. All of the copper-rock taken out was classed as stamp-rock.

The mineral land was heavily timbered, mainly with hard-wood. The Copper Range Railway, who constructed their road in 1899, passed within half a mile of the Winona workings. The Copper Range did not begin hauling freight until December 1, 1899; prior to that time all supplies and machinery were hauled either from the Greenland station, on the Chicago, Milwaukee and St. Paul Railway, 16 miles in distance, or from Houghton which was 27 miles away. During the winter the shorter haul from the Greenland station was available, but later the road became impassable and the Houghton Station had to be used as a base. Little construction had been completed, since this Company was using existing structures built in 1864 and later.

During 1900, the operations were carried on mainly for the purpose of testing the most promising portions of the amygdaloid belt which they had been developing, with the hope of finding sufficient pay ground to warrant the erection of a stamp mill. They had their ore stamped at the Atlantic Mining Company's mill, and would change to the Baltic mill when the Baltic could make room for Winona's ore.

The No. 1 shaft was now 8 by 18 feet and 410 feet deep. Shaft No. 2 was also 8 by 18 feet, but 375 feet deep. They were cross-cutting west from the bottom of the No. 3 shaft, which was 16 feet deep.

The only construction done during 1900 was building a small ice house and meat market, and the rebuilding of the store which was burned in February of 1900. The annual report was signed by F. W. Denton, Agent.

A year later, 1901, President Stanton reported "The operations during the year include the completion of the diamond drill exploration across our lands, which, however, failed to reveal the existence of any deposit carrying copper enough to be of value". The total drilling in rock was 2,889 feet in 5 holes and the total length of stand pipe was 235 feet. These tests did not seem to warrant the expenditure involved in the construction of a stamp mill, and additions to the mine plant would be required in order to handle a large quantity of rock. They only worked in the No. 2 shaft during 1901, with a total sinking of 298 feet. Neither the report nor the statement of working expenses showed any construct-ion this year.

President John Stanton reported that during 1902 the underground work was again confined to drifting on the lode in the direction in which the previous work had indicated the probability of discovering paying ground, and in making preparation for a mill test.

The only stampmill that was available was the head at the Atlantic Mine Company's mill which was also engaged in stamping rock from the Champion Mine. The head was not ready for Winona's use until December 3rd, when 3,772 tons of ore, or 92 railroad cars of rock, were shipped and stamped. Freight costs to the Copper Range Railroad for hauling these 92 car loads, was $79.64. The stamping reports were quite satisfactory.

All of the rock sent to the mill was obtained from the 2nd, 3rd and 6th levels, with the exception of 103 tons which came from the stockpile on the surface. It may be well to state that there was no first level in this shaft and that the 5th was not opened when the shaft was sunk. Therefore the stamp rock all came from levels opened to the south of the No. 2 shaft. A map of the undergound shows that the 2nd level connects shaft No. 1 and No. 2.

On the surface, considerable work was necessary to prepare for shipping rock. A spur 3,600 feet long was built by the Copper Range Railroad to the shaft. The roadbed was built by the Winona Company, and the track-laying and surface were done by the Railroad Company at their expense.

A new 18 by 24 foot rock-breaker was purchased and set in position over the railway track so that the crushed rock fell from the breaker into the cars. A second-hand slide valve engine was purchased for driving the crusher, and a new 3 inch steam line about 400 feet long was built to carry steam to this engine.

On February 10, 1903, a reporter from the Native Copper Times of Lake Linden wrote: "The yield of mineral from the Winona rock stamped at the Atlantic stamp mill has been a surprise to everybody. With only a very moderate amount discarded, the yield was about 40 pounds to the ton stamped which is quite extraordinary and establishes a new value for the mine. And the test is said to have been a very fair one and made for the purpose of learning the actual mineral value of the lode rather than to make any spectacular show. The results were so satisfactory that the company has decided to increase the mine force and begin developing the lode on quite a liberal scale. Six or seven new power drills will be started up almost immediately and sinking begun for another shaft, besides extending the present openings. When it is remembered that the yield of mineral per ton is no larger at the Trimountain nor Champion than that stamped from the Winona, then the significance of tests made may be fully realized."

During 1903 the Company employing 125 men sunk another shaft for the purpose of ventilation, as well for hoisting rock. This shaft, No. 3, was started 1,536 feet south of the No. 2, and was made by sinking from the surface, at the same time rising from the 4th level. Due to the late disappearance of snow and excessive rains, they were not able to start shaft No. 3 until June, while raising from the 4th level began in April.

This shaft was 6 by 18 feet inside timbers and had two 6 by 6 foot skip compartments, and a ladder and pipe road at one end. The alignment of the shaft was perfect, and the work was a credit to Mr. W. W. Stockley who did the surveying, and to Captain John Peterson who carried the lines so well. The equipment for the No. 3 shaft consisted of the small single drum geared hoist formerly used at the No. 1 shaft, and one boiler from the early No. 1 plant. The log house built while the exploring pit was being sunk in 1901 was made to serve as a boiler and engine house for the new shaft. The hoist operated a small skip holding about one ton.

A small change-house had been erected near the No. 3 shaft and was supplied with steam heat. A temporary shaft house had also been erected over the shaft and a three inch air pipe had been laid between the No. 2 and 3 shafts, and a 5 inch air line from the compressor to the No. 2 shaft replaced the old 3 inch line. $622.11 had been spent installing pipe connections for air and steam with the crusher for the No. 3 shaft. The Company had also built a new road around the No. 2 shaft to the saw mill, which now connected the branch road to the No. 3 shaft.

$21,236.74 was spent on the No. 3 shaft.

The Winona also had a sawmill 30 by 70 feet with a 42 by 66 foot wing for boilers, a 16 by 46 foot wing containing a shingle mill, and a plant having a daily capacity of 20,000 feet of sawed lumber. The main line of the Copper Range Railroad passed alongside the sawmill.

During the year, two more double dwellings were erected, giving accommodations for four more families. The office, which was a log building, was covered with siding and painted. $2,470.00 was spent on the contract for the new dwellings.

The year 1904 found the officers of the Company much happier. The test made by mining and stamping a total of 89,237 tons of rock, which yielded 1,784,157 pounds of refined copper, or an average of 20 pounds to the ton, pointed to the conclusion that if a sufficient extent of that kind of ground could be developed to admit an output of 1,000 tons of rock daily, the expenditure necessary for the erection of a stamp mill would be fully warranted. 45,115 pounds of copper mass were shipped directly from the mine to the smelter.

The mining agent reported that the copper ground was more or less bunchy and was separated by bars of barren vein or trap. This fact led to the "filling method" in working the stopes. By using this method the stopes were filled with the barren rock instead of hoisting the entire rock blasted indiscriminately.

The small hoist of the No. 2 shaft had reached the limit of its capacity, and the still smaller hoist at the No. 3 shaft was inadequate for the new work. Accordingly, larger hoists were purchased capable of hoisting a net load of 1 1/2 tons from a depth of 1,500 feet. The hoist at the No. 2 shaft was a second-hand hoist obtained from the Trimountain Mining Company, and that at the No. 3 was shaft a new one obtained from the Nordberg Manufacturing Company. Both were geared hoists. A combined engine and boiler house was erected at the No. 3 shaft. The building was 34 by 48 feet and was sheathed outside with corrugated iron.

According to the 1905 annual report, which ended in December of 1905 but published in March of 1906, Company President John Stanton died on February 23, 1906. He had been with the Company since its organization and the new President was Arthur G. Stanwood. The Officers and Directors also thought the time had come for a change of policy. The ground near the No. 2 shaft had continued to be disappointing, but that around the No. 3 appeared to be uniformly good and they would apply their efforts to develop this ground. Not much construction was completed during 1905, according to Agent F. W. Denton. Both hoists were operating satisfactorily and the new barn, which was destroyed by fire in December of 1905, was erected.

The Winona Copper Company's eighth annual report, for 1906, showed some changes in its leadership. Arthur G. Stanwood was still the President, Nathaniel H. Stone, Vice-President, and Edward B. O'Connor, Treasurer and Secretary. Directors were Arthur G. Stanwood, Charles J. Paine, George P. Gardner, William A. Paine, Nathaniel H. Stone, Walter Hunnewell and James H. Seager.

The annual report showed that the stockholders were assessed five times by this date. Their first was for $265.00; second raised $592.00; third brought in $3,472.00; fourth added $98,828.00 to the treasurey and the final added $191,714.00 to the Company's funds. The new management had lots of plans for the next year, all very expensive.

General Manager Lucius L. Hubbard appointed Rex R. Seeber Superintendent, who took charge of the operations of the mine as well as the construction work. The miners were using 12 air-drills and development work continued in the No. 3 shaft. They had started on the No. 4 shaft and the shaft house was built and in use by autumn of that year. Shipments of rock to the Adventure Mill began on the completion of Winona's steel shaft and rock house in the middle of October, and had continued.

During 1906, at the No. 3 shaft a modern steel shaft and rock house were erected by the Wisconsin Bridge and Iron Company. It was about 28 by 71 feet, and 49 by 39 feet, respectively, 87 feet high to center of sheave, 100 feet over all, and covered with galvanized iron. It had a bin capacity of 700 to 800 tons. It was provided with two 18 by 24 inch Blake rock-breakers and a drop hammer. All were operated by a small steam engine. A poor-rock trestle connected this building with the waste pile. Building the shaft and rock house cost $32,875.49, with $1,494.77 being spent for the hoist.

Besides the railroad spur being added to, a coal trestle 234 feet long was built east of the No. 3 shaft, with an average height of 17 feet, for $5,208.45. It connected with the new railroad spur, and the coal bins had a storage capacity of 3,000 tons.

North of the coal bin was the new steel-roofed, brick boiler house, 50 by 66 feet, containing one 268 horse-power Parker boiler. During the next year some of the old boilers from the No. 2 compressor and hoisting plant would also be moved there. Construction of the No. 3 boiler house cost $8,779.21.

The new power house was erected just east of the boiler house and was a steel-roofed, brick structure, 58 by 85 feet, with a traveling crane capable of lifting 7 tons. In it, erected and ready to use, was a 16-inch and 30 inch by 36 inch cross-compound, condensing Allis-Chalmers engine, with a 250 K.W. generator, and the necessary components of an electric plant. In this same building was an 18 inch and 32 inch by 42 inch cross-compound condensing Nordberg air-compressor, that had a capacity of 2,700 cubic feet of air per minutes at 80 pounds pressure, and which could operate 40 drills. In connection with it and the electric engine, a Worthington condensor was installed. There was still vacant floor space in the power house for another large electrical unit to be installed in the near future.

A water main, 6 inches in diameter, was laid 8,050 feet from the Sleeping River across the King Philip location, to a steel tank of 60,000 gallons capacity. This stood on the high ground southeast of the No. 2 engine house, and provided water for the power plants on each location, as well as for some domestic use. The new water system cost the Company $7,618.59.

During this same year, 1906, a substantial electric light and power line had been built 6,200 feet through the location and west to the King Philip at the cost of $16,406.68. Telephone wires also connected the important parts of the mine and surface structures.

A new wood-frame office and warehouse building had been built south of the No. 3 shaft. The office was two stories high, 24 by 37 feet, and had a stone basement; a vault was built in the basement and a space left for its upward extension. The warehouse was 20 by 40 feet.

South of and near the new office was the machine and blacksmith shop. The former was 50 by 30 feet and the latter 40 by 30 feet, both under one roof. The machine shop had a seven foot stone basement. A drill-press, lathe, planer, pipe-cutting tools, bolt and nut machine, grind stone, etc., were also installed. The blacksmith shop had a drill-sharpening machine and a foundation for a steam hammer.

The carpenter shop was 35 by 60 feet, had a stone foundation and was a wood-frame structure. It had the necessary tools for keeping the copper mining industry going.

A new dispensary and three new boarding houses were built on the location, as mentioned earlier. The Superintendent's resident was a wood-frame, two-story building, 29 feet by 37 feet. Mr. Rex R. Seeber now lived on the east side of the Winona property, with residental blocks 10, 11, 12, 13, 14, 15, and the Company store, just west of his new home.

Three additions were made to the No. 3 dry house, 24 by 24 feet, and 12 by 26 feet, and 10 by 22 feet. Two sheds for temporary use were also erected, 12 by 24 feet and 16 by 20 feet. At No. 4 a temporary shaft house, 20 by 48 feet, was in use, and also an engine house, 14 by 28 feet. A pump house was built by the Winona at Sleeping River, on the King Philip location. It was 22 by 30 feet.

The Company sustained a severe loss in the death of its President, Mr. Arthur G. Stanwood, on the 30th of December 1907. Charles J. Paine Jr., was elected a Director, to fill the vacancy.

The 1907 annual report was signed by Nathaniel H. Stone, Vice-President, and he noted that development underground had progressed at the mine, and shipments of rock to the Adventure stamp-mill were continued. They reached a maximum of 500 tons, and averaged about 400 tons daily until November of 1907, when, owing to the state of the copper market, rock shipments were discontinued. At that date the working force was reduced, and attention was directed to the opening of new ground. All of the copper ore had come from shaft No. 3. In the

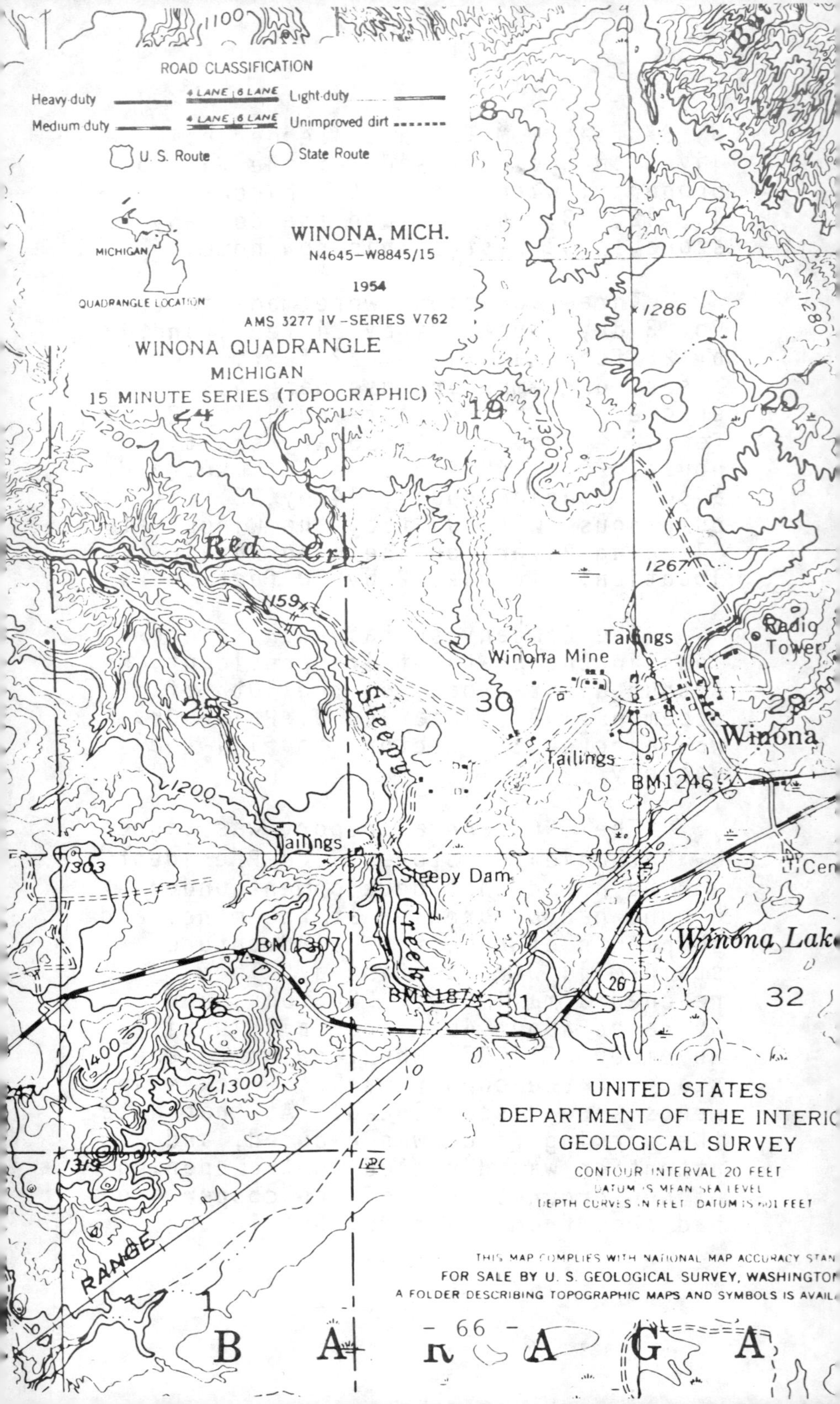
ROAD CLASSIFICATION
Heavy-duty
4 LANE 6 LANE
Light-duty
Medium-duty
4 LANE 6 LANE
Unimproved dirt
U. S. Route
State Route
MICHIGAN
QUADRANGLE LOCATION
WINONA, MICH.
N4645—W8845/15
1954
AMS 3277 IV—SERIES V762
WINONA QUADRANGLE
MICHIGAN
15 MINUTE SERIES (TOPOGRAPHIC)
1100
1200
1286
1280
1300
Red Cr
1159
1267
Tailings
Radio Tower
Winona Mine
Sleepy
Winona
Tailings
BM1246
1200
Tailings
1303
Sleepy Dam
Creek
Winona Lake
BM 1307
BM 1187
26
1400
1300
1319
RANGE
UNITED STATES
DEPARTMENT OF THE INTERIO
GEOLOGICAL SURVEY
CONTOUR INTERVAL 20 FEET
DATUM IS MEAN SEA LEVEL
THIS MAP COMPLIES WITH NATIONAL MAP ACCURACY STAN
FOR SALE BY U. S. GEOLOGICAL SURVEY, WASHINGTO
A FOLDER DESCRIBING TOPOGRAPHIC MAPS AND SYMBOLS IS AVAIL
B A R A G A

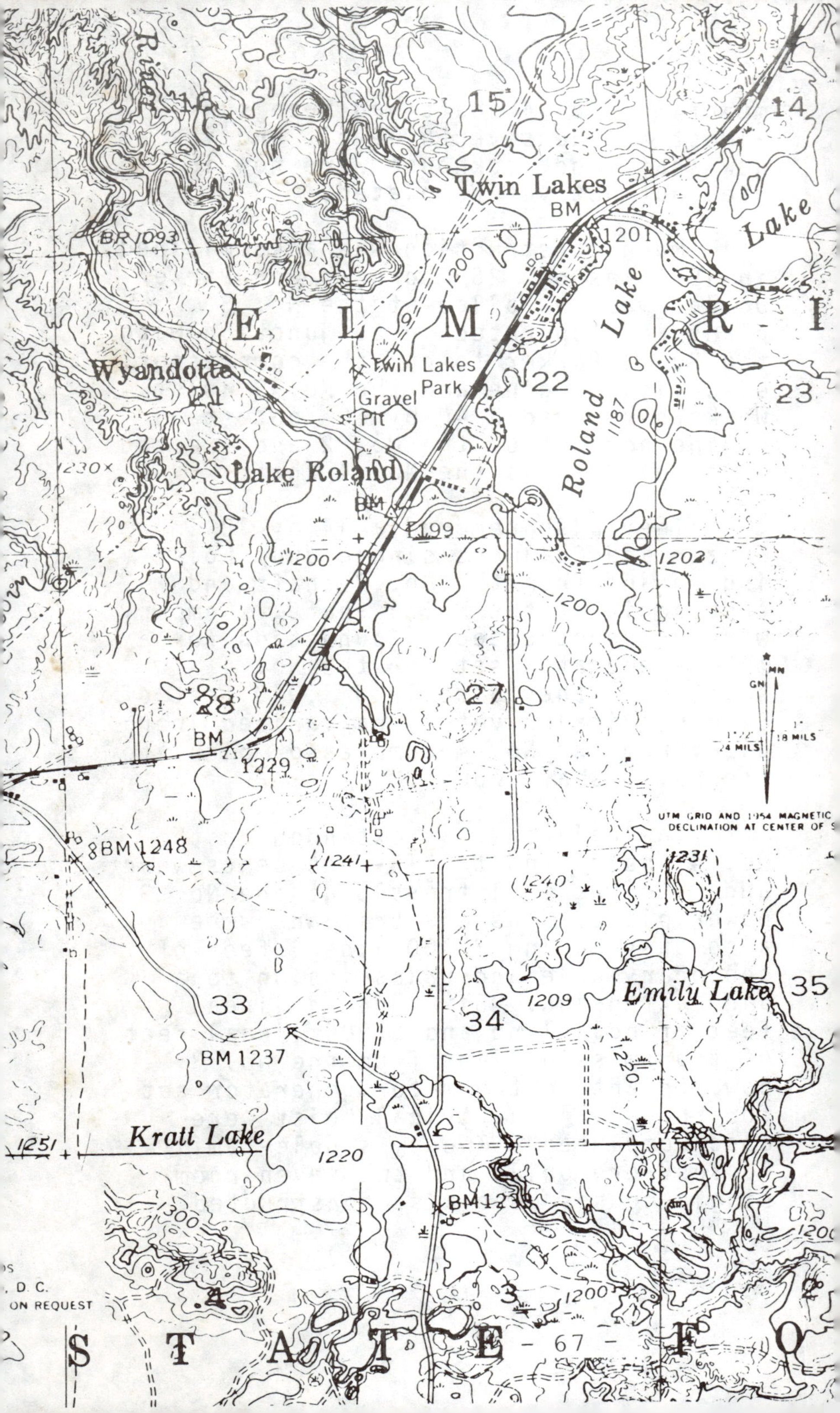

15
14
Twin Lakes
BM
1201
Lake
BR 1093
E L M
R I
Wyandotte
21
Twin Lakes Park
Gravel Pit
22
Roland Lake
23
Lake Roland
BM
1199
1200
1202
27
28
BM
1229
BM 1248
1241
1240
1231
UTM GRID AND 1954 MAGNETIC
DECLINATION AT CENTER OF S
GN
MN
24 MILS
18 MILS
1209
Emily Lake
35
33
34
BM 1237
Kratt Lake
1251
1220
1300
1200
D. C.
ON REQUEST
S T A T E
F O

meantime, No. 4 shaft had been sunk almost to the seventh level, and its new shaft and rockhouse were now in use, but not yet completed.

Much construction was accomplished in 1907, as one 268 horse-power Parker boiler was installed at the No. 3 plant, and operated in conjunction with the No. 1 Parker to run the compressor, electrical engine and the No. 3 hoist. Workers removed a 40 by 40 foot steel engine house from the No. 2 and erected this building at the No. 4 shaft.

They also removed a 14 by 18 inch Frazer and Chalmers single drum hoisting engine from the No. 2 shaft and erected it at the No. 4 shaft. Company workmen constructed a foundation for a motor-generator set for the No. 4 shaft, and they also laid 1,910 feet of 8-inch spiral riveted, galvanized iron pipe from the No. 4 compressor plant to the No. 4 shaft-house.

The railroad was extended to the No. 4 shaft, and two 15-foot bents were added to the coal trestle at the No. 3 power plant. The electricians were also busy adding 2,100 lineal feet of secondary pole and transmission to a line extension. Six hundred lineal feet of pole line and 2,900 lineal feet of transmission line from the No. 3 power plant to the motor-generator set for the No. 4 electrical hoist were also completed. Twenty single-family houses were built and two seven-room single houses were also constructed.

The 1908 annual report showed a capital stock of one hundred thousand shares at par $25.00, with $17.00 per share paid in. Mr. Charles J. Paine, Jr., filled the vacancy of President on June 22, 1908. During this year the price of copper remained so low that it was considered inadvisable to renew the rock shipments which had been discontinued in the autumn of 1907. Underground development had been actively pursued and the new openings had been very encouraging, especially those tributary to the No. 4 shaft, where the showing was the best in the mine. The No. 3 shaft was cut open to the thirteenth level, plat cut and skip and ladder roads completed, and all levels made ready for renewed production. After all of this the shaft was closed down.

The rock-house at the No. 4 shaft had been completed, and if shipments were resumed in the spring the two shafts combined would produce 750 tons daily, and more than that amount with increased development.

As with the King Philip Board of Directors, the Winona Directors had been considering for some time the erection of a stamp-mill on or near the mining location. Careful measurements had been made for a period of about two years and the water flow of the Sleeping River was considered capable of handling a four-head stamp-mill.

The King Philip Copper Company was ready to join with the Winona and pay half the cost of the proposed mill.

General Manager Lucius L. Hubbard had recently estimated that the Winona Mine had enough mining rock to keep their half of the proposed mill busy for about six years. The saving in transportation and stamping charges by treating this rock in their own mill, near their location, was estimated at not less than 20 cents per ton. During the February 3, 1909 annual meeting, the Directors voted to have plans and specifications drawn up for a two-head stamp-mill, and appointed a committee with power to accept the same in conjunction with a similar committee appointed by the King Philip Copper Company.

The principal items of construction completed during 1908, were the rock-house at the No. 4 shaft, with crusher, motor, rock trestles, wiring for light, power and telephone, and pulley stands to the engine house. Also the installation at the No. 4 hoisting plant, of the motor generator set, switchboard and direct current motor geared to the hoist: also lining the building with brick and providing a small boiler for heating purposes when the plant was shut down.

Company workers completed some alterations in the warehouse, and the rebuilding of the machine and blacksmith shop.

Development work had been confined to the No. 4 shaft during 1909, where eight drills were kept at work. It was sunk to the 12th level, the shaft was cased and the ladder road and 5-inch air line were brought down to that point.

Aside some small expenditures made for No. 3 rock house and the No. 2 hoisting plant, the principal items of construction were as follows: No. 4 rock house - a drop hammer, 24-inch by 36 inch crusher, and rock bin with appurtenances were installed. Employees strung 3,000 feet of telephone wire and 6,900 feet of electrical wire. At the blacksmith shop, the old No. 2 steel boiler house was moved next to the machine shop onto a stone foundation, and enlarged to 65 by 41 feet, to be used as a blacksmith shop. Foundations were made for the necessary tools. As far as the water supply, 1,140 feet of permanent water lines were laid. Several more homes were constructed, one each for the mining captain, mine physician, and stamp-mill superintendent.

According to the General Manager, the excavation for foundation of the two-head stamp mill commenced on the 19th of May, 1909, and was practically completed by July. In July they began to sink the caissons for the stamp heads. A railroad track was graded and laid from the main spur to the head of the mill for the delivery of rock, sand, cement, and other material. The excavation for the concrete core at the dam was also finished and the laying of the core began.

The mill secured its water from the Sleeping River by a dam, with a storage capacity of 150,000,000 gallons. Careful hydrographic surveys for three years showed that the watershed tributary to this river yielded about 3,000,000 gallons per day, which was enough for the mill. The dam impounded sufficient water to operate the mill for two months. Early in September both caissons were bottomed at about 57 feet on stratum of clay, and the caissons were filled with concrete. Early in December all the foundations for the mill and coal trestle were completed and the boiler house and smokestack erected.

The annual report for 1910 found President Paine and the Directors informing their stockholders that the Winona property was bounded on the southwest by the King Philip Copper Company, and it would be a decided advantage to both properties to be merged. The Winona had a materially larger investment, both underground and on the surface, than the King Philip. On the other hand, the King Philip had about 25 percent more ground which they believed carried as much copper as that portion of Winona ground between the southwest boundary of the Winona property and a point about one-half way

WINONA photos often have been used in the weekend section to detail the stories about this small Houghton County community which is off the beaten path so that the motorist moving between Houghton and Ontonagon seldom turns to the right to see what is left of the community. Here is the No. 4 shafthouse of the Winona Copper Co. which maintained a productive existence until slightly after the end of World War I. (Ernest Kuopus photos)

between the Winona No. 3 and No. 4 shafts.

Northeast of this point on the Winona property, it had been pretty well demonstrated that there were no values near the surface. On the southwestern part of the King Philip property there had been no explorations whatsoever. A vote would be taken at the next Winona and King Philip annual meetings.

The number of drills employed during 1910 varied between five and eight, being six in December. From time to time drills not needed at the Winona were transferred to the King Philip. Owing to the age of the timbers in the openings of No. 3 shaft and the tendency of ground to cave in, these timbers needed immediate attention, and it was considered to be a wiser policy to draw on the sawmill supply in the near future. The sawmill was located on the Winona property, alongside the Copper Range Railroad spur and near the Winona Creek. It was by the No. 3 shaft and hoist area.

In connection with the building of the stamp mill, there was installed at the No. 3 Winona power house, a steam turbine of the horizontal Curtiss type, rated at 750 kw. maximum, at 80 percent power factor. It had a Worthington condenser directly under it, connected to it by a single sheet copper expansion joint. A 150 kw. motor generator for operating the locomotive took an alternating current at 2,200 volts, and delivered direct current at 500 volts. These additions to the power plant rendered necessary the installations of one more water tube boiler of 303 h.p. to carry steam at 175 pounds.

Under expenditures at the mine for the year 1910 were included expenses for a new office building, a fire department, quarrying, machine shop construction, blacksmith shop construction, electric light and telephone extensions, power plant overhauling, round house construction, steam plant overhauling, oil house construction, No. 3 rock house remodeling and air line construction.

The stamp mill, after many delays, was nearing completion. Each head was expected to crush 700 tons or more per day. A coal bin and boiler house immediately adjoined the mill on the east, at the nearest point to the heads. All other mechanisms in the mill were operated by electricity from the generators at the No. 3 shaft power plant.

The principal difference of operation between this stamp mill and other Lake Superior mills was in the re-use of a large part of the wash water through the medium of six large settling tanks located under the washfloor. The Winona had estimated that they had a water supply sufficient for four heads. The sand-tailings were conveyed on a belt to a distance of 321 feet from the mill and 124 feet above the ground. The slimes were delivered by a steel launder 1,200 feet long to a ravine beyond the sand discharge. All the rock would be hauled from the shafts to the mill by an electric locomotive. Rex R. Seeber was still the superintendent, Allan Eddyvean Mill Superintendent, John Broan the mining captain and Frank Lukey, clerk.

The Officers and Directors of the Winona acquired the King Philip Copper Company in May of 1911. During the Winona's annual meeting held on March 28, 1911, the stockholders agreeded to increase the shares of their company by 100,000, and offer to exchange 66,666 of the same for shares of the King Philip Copper Company at the rate of two Winona Shares for three King Philip shares. This would leave 33,334 shares in the Winona treasury. Stockholders who did not attend, voted by proxy.

During 1911, only the No. 4 Winona shaft and the No. 1 King Philip shafts, which were side by side, were worked. At the 12th level the Winona No. 4 shaft was connected to the No. 1 King Philip 8th level.

The stamp mill, as anticipated, went into operation in March of 1911. Since that date, they had were stamping rock from the No. 4 Winona shaft and the No. 1 King Philip, and could take care of more rock than was furnished from the drills in commission. The second head would go into commission as soon as the necessary addition to the underground work force could be obtained, as the management suffered from a shortage of labor. The Company believed that greater economy and a large amount of rock could be furnished by temporarily closing down the No. 1 King Philip and opening up No. 3 Winona shaft.

Two Hardinge tube mills had been added to the one already in operation; one six feet in diameter, with a five foot straight face, and the other eight feet in diameter with an 18 inch straight face, geared to and driven respectively by a 35 h.p. and a 45 h.p., 3 phase induction motor.

The one large obstacle the Company encountered during 1912 was the shortage of trammers. Notwithstanding the fact that the Winona paid higher wages than any other mine in the district, it was difficult to get the unskilled trammers to work at a mine so far from moving pictures and electric lights of the larger towns and cities. They started to look for extra men during January of 1912 to increase rock output, but gained few men until spring when the breaking up of the lumber camps resulted in a number of men entering towns and seeking employment in the mines. However, when winter started and logging camps opened, the Company lost them again.

In the late summer they gained a few men, but as winter came on the movement was against the Winona, and for December, 1912, in spite of continous effort on their part, the average number of underground men was almost exactly the same as for January, 1912. Winona shafts No. 3 and 4, and King Philip shaft No. 1 were worked during the year.

The stamp mill had operated during the year with one head almost continously in operation, and two heads working part of the time. Like other mines in the district, they had suffered from lack of trammers, and had resorted to every available means to hire more men. The inevitable results of insufficient trammers were, first, a falling off in the tonnage stamped, and, second, a decrease in the percentage of copper per ton recovered, because in the effort to make good the tonnage, rock of inferior quality inevitably went to the mill. It was because of this, in part at least, that the percentage of copper recovered in 1912 was below that of 1911.

Early in the year the stamp mill was examined by an expert, who commended the general plan of construction and operation, but recommended the addition of further regrinding apparatus. This was ordered to the extent of three Hardinge tube mills, which were attached to head No. 2 and could be used in parallel when both heads were running; or in tandem, when No. 2 head was not in commission, on material from No. 1 that would otherwise be discarded.

President Charles J. Paine Jr., wrote that the beginning of the year 1913 found the Company struggling against a shortage of labor. The report was unquestionably a discouraging presentation of the situation. He made no attempt to gloss over any of the difficulties which worked against the mine. As the months passed by, this shortage became more acute and came to a climax on July 23, 1913, when there was a strike of the miners of the district under the leadership of officials of the Western Federation of Miners. Practically all the Companies in the Copper Country district were obliged to stop mining and in some cases were unable even to keep the water pumped from their mine shafts. Much disturbance and rioting took place.

The main point at issue was the recognition of the Western Federation of Miners. The record of strikes in the past to which this Union had been a party had been such that it was a foregone conclusion that neither the mine managers, mine directors, nor stockholders would for an instant consider having any dealings with the Union and there was never a possibility of anything but defeat for the Union. The strike wsd not called off and many special watchmen, etc., had to be employed by the mines, but the indications are that it ceased to be an important factor in the situation and would gradually die. It had been kept alive to a large extent by political investigations and would-be arbitrators.

The Winona did not attempt to resume mining until October of 1913, and for many weeks thereafter only a very small rock output was possible. Toward the end of the year the work force began to increase and this tendency continued. For a year and a half before the strike the labor shortage was confined to trammers. Now there were plenty of trammers, and what shortage existed was of miners, and that shortage was soon to be relieved. The labor situation looked better than for the last couple of years, the indications that all of the men that could be set to work would be available.

It was needless to say that the interruption to operations caused by the strike had been very costly to the Company, yet the final result would insure a renewal for a number of years of industrial peace and prosperity in the district and would be a great benefit to all concerned. The cost for special watchmen and incidentals cost the Company $10,966.46 that year.

Superintendent Rex R. Seeber reported that on July 23rd, notice of a strike was posted on the door of one of the change-houses. One man stayed home. At noon, two more men came up from the mine. The night shift went down at both shafts to about one-half the usual number of men. On the 24th, No. 3 shaft only worked on the day shift, and there were not sufficient men on the night shift to work the shaft.

Mr. Seeber said that from all informtion obtainable, it appeared that very few of the workers joined the federation until after the strike was called. Order was preserved on the location and no arrests were necessary on account of the strike. On October 6th, operations were resumed on the day shift at the No. 3 shaft. They started with only seven men underground, and at the end of the year had over seventy men working underground. Common labor was now plentiful, but the Company had not yet obtained sufficient miners to start a second shift.

Apparently Mr. Seeber did not know quite what was happening on his property, as the front page of the Calumet News reported "Deputies Sent to Winona Location." Seems that a disturbance had occurred early the morning of December 18th, when several volleys of rifle shots were fired, apparently to scare non-union workmen. A squad of Houghton County deputies was sent to the scene and remained for an extended period. The source of the shots was unknown. With the additional force that was ordered to Winona, Houghton County was now patrolled by 2,100 deputies and every locality from Toivola to Copper City was controlled by civil authorities.

It was also reported in the Miner's Bulletin that on Friday afternoon Sheriff Cruse and deputies went to Winona and threw John Merkle, a striker, his wife and small child out of one of the company houses of the Winona Mining Company. These people were driven out in a snowstorm and their furniture thrown into a snow drift. This act was done without due process of law, it is stated. A neighbor of the Merkles' went to the woods, gathered some fuel and built a fire for the ejected family that they might not freeze before they could be taken care of. This neighbor was arrested for trespassing, brought to Houghton and lodged in jail. It is also stated that the sheriff notified several other strikers who occupied company houses that they would be given the same treatment unless they vacated their homes at once. The *Miners Bulletin* was known to exaggerate a bit at times.

In June of 1913, Mr. Alton L. Dickerman, Mining Engineer, was employed to make a study of the extraction problem in the mill. His work had been interrupted by the strike, but was completed by the end of the year. His report was added to the annual report to the stockholders for 1913.

The Sixteenth Annual Report of the Winona Copper Company indicated that in the early months of 1914 conditions were slowly becoming normal as the miner's strike gradually died out and by July the Company had reached the maximum output of its career.

Winona, Michigan, from left to right, Roger Maki, Sherwood Maki, Alberta Maki, Deland Harma and Elmer Maki leaning on the car. Courtesy of the Harold H. Heikkinen collection, Atlantic Mine, Michigan

The regrinding department in the mill was aiding materially in the saving of finely disseminated copper in the rock. The cost of producing copper was slowly lowering month by month until the European war broke out and caused so general an upset to the copper business that it was thought advisable to completely shut down the mine. Orders to that effect were issued on August 6th.

On October 15th, the Company authorized leasers to take over and production was resumed. Thus some thirty-five of the best of the former employees were able to work certain stopes on the third, fourth and fifth levels from the No. 3 shaft.

The Company benefited indirectly in several ways, such as by being able to dispose of sundry perishable mining supplies on hand and getting rent from more of the Company houses, besides receiving half of any profits from the operations after allowing the tributors a very moderate rate of wages. Moreover, by this arrangement the Company was able to keep a number of extra good workmen on the location, according to President Charles J. Paine, Jr. The Company still had $45,310.45 of cash in the bank, $47,979.63 in supplies, $15,133.03 in accounts receivable with 580,643 pounds of unsold copper.

The only construction work of the year was in connection with the experimental work on the Shields Classifier and Lovett Grinding machine.

The year of 1915's annual report told of the tributing or leasing arrangement outlined in the past annual report ran from October 15, 1914 to June 1, 1915, with satisfactory results to the Company as well as the tributors. The indirect benefits from it included that the Company's share of the profits was approximately equal to the taxes and other necessary expenses to operate during the life of the leasing arrangement.

During June of 1915, the Keweenaw Miner reported that Captain J. M. Broan, who had held the position as head mining captain of the Winona, had tendered his resignation to the Company and had moved to Houghton. He was succeeded by Captain John Pooley of the Houghton Copper Company. Both men were experienced mining men and were old partners, having worked together in the old Atlantic Mine.

The King Philip No. 1 shaft house which had been destroyed by fire a few years before, was replaced. The Winona had collected $1,000.00 in insurance. It would shortly increase the output some 35 per cent.

Work at the mine was resumed early in May when unwatering of the No. 4 shaft was commenced. Some copper was produced in June, and production had gradually increased. An electric pump, 4 1/2 by 8 inch triplex, single acting with a 35 h.p. motor, 600 rpm, 2,200 volts, was installed at the ninth level of the No. 4 shaft. This pump had been the principal means of unwatering and would ultimately be installed at the twelfth level of No. 4 shaft to handle practically all the water from the mine.

President Paine reported that operations during 1916 had been greatly hampered much of the time by a labor shortage. Expenses for labor, coal, powder and supplies of all kinds had increased greatly and were reflected in the cost of producing copper. Copper sold at 28.03 cents per pound.

During 1916, Winona shafts 3 and 4 were operated as well as King Philip shaft No. 1. The King Philip showed the best. There were 225 men working underground and at least 100 more men would be needed soon. By increasing the size of the motor driving the Hardinge Mills and using steel balls instead of pebbles, they had been able to materially increase the amount ground in each mill. These changes together with rearrangement of mineral bins, pumps and outside slime launders in order to handle custom stamping, were the only construction items in the mill for the year.

During the year, a wooden shaft and rock house was constructed at the No. 1 shaft of King Philip. This rock house contained an 18 by 24 Hodge crusher driven by a 50 h.p. 2,200 volt electric motor. A mass hammer type was operated by air. To equip the No. 1

October 9, 1988

King Philip Shaft No. 2
October 9. 1988

shaft for regular operation, stringers were put in the shaft and considerable work was done at each of the levels in retimbering, according to Rex R. Seeber, Superintendent.

For about nine months of 1917, mining operations were carried on by the Winona. Throughout that period, there was a marked tendency for costs of all kinds to increase, coupled with the lack of adequate labor, though for six months of that time the Company was able to show a small margin of profit. Beginning in July the underground openings ran into much poorer ground, the price of copper was sagging, and when the results of the month's operations were obtained it became evident that the mine would have to be shut down.

At that time the fixing of a price for copper by the Government was pending and it was thought best to await the announcement before giving orders to stop. Immediately upon the announcement by the National Government of a price of twenty-three and one-half cents, orders were given to shut down the mine.

Superintendent Rex R. Seeber recommended a continuance of operations on a somewhat smaller scale and only in certain parts of the mine. It did not seem advisable to the Board of Directors to follow his recommendation. Whereupon he made an application to work the property on tribute, guaranteeing a small minimum payment per pound of copper produced and an additional payment dependent on the price at which the copper was sold exceeding a minimum figure. The Winona property had been operated by Mr. Seeber on this basis since October 1, 1917, and with satisfactory results, according to President Paine.

According to Superintendent Seeber the stopes opened showed about the usual run of ground but the shortage of labor and its inefficiency after June 1917, dropped the grade of the rock mined very seriously. There was no new construction work for the year. After October 1st, some rails, pipes, etc., were removed from the No. 3 shaft.

The Winona Copper Company started a second and new life, according to the twentieth annual report for the year of 1918. There was a complete change of Officers and Directors. The President was now Thomas S. Woods, while N. Wheaton Dean was the Vice-President and Arthur C. Paine was the Treasurer and Secretary. Directors were Thomas S. Woods, Patrick D. Burke, Alfred B. White, N. Wheaton Dean, John R. Findley, Richard H. Summers and Gustav A. Braun. The Company had a capital stock of $5,000,000 of 200,000 shares, 166,667 shares issued, fully paid.

President Woods told the stockholders that the new Board of Directors was elected on April 16, 1918 they had given much time and study to the affairs of the Company, and as a result suggestions were given to the stockholders for consideration.

After organization, the Directors removed the Boston office of the Company to 60 Congress Street. They continued the former management's policy of keeping eastern salaries and expenses within modest limits. In May of 1918, after two officers had visited Winona and examined the mine and mill, operations on Company account were resumed with the lessee Mr. Rex R. Seeber again in charge as general manager. Production had been small and costs consequently high.

This annual report continued to give the stockholders a general historical view of the Company. The Isle Royale Company, the one successful producer operating on the Winona-Isle Royale lode, developed good ground in their No. 2 shaft only below the 25th level, although other shafts further south were more fortunate. This mine showed high costs and deficits until production had been brought above 1,200 tons per day of rock yielding about as much copper as Winona's own.

Convinced that Winona could become a profitable producer only by following the methods through which Isle Royale had achieved success, the new Winona Directors had equipped the hoist at King Philip No. 2 shaft for deeper mining, believing with many Michigan

One of the two pictures painted by Flora M. Knudsen's art class that hangs in the school

men, that the south ground gave promise of better copper at depth.

Improved values had to be found and extensive new openings made to furnish an output large enough to insure economical operations. At the mill, which required few changes to permit the efficient treatment of over 1,250 tons daily, an oil flotation plant for saving the fine copper which was being lost in the tailings, was now installed and ready to operate.

Because of market conditions, the mine was closed temporarily early in February of 1918. During the suspension, the new superintendent, G. A. Braun, who took charge on November 15th, 1918, worked out a plan for consolidating the widely separated power plants of the two original mines on an economical basis. In January, 1918, the Company's options on valuable timber south of Winona were renewed for a term of ten years, through April 1, 1930.

Superintendent Braun's part of the report stated that it was on May 1st that the Winona assumed and continued operations for the balance of the year, yielding fine copper production for the eight months of 819,857 pounds, making a total of 1,576,683 pounds of ingot copper for the year. He also wrote that a one-story concrete block building, 48 by 66 feet with a wing 22 feet by 35 1/2 feet was built adjacent to the northwest corner of the stamp mill to be used as a flotation building. In November all construction at the flotation plant was suspended due to high supply cost and wages and scarcity of labor.

Under orders of the Directors, careful inventories of all assets of the Company had been taken, and all books and accounts readjusted to meet present day requirements. The revised figures were presented to the stockholders and were certified by the Company's Auditors.

President Woods reminded the stockholders that Winona was not yet a mine but was rather a large prospect with costly surface equipment and mill, besides houses and timberlands, all together worth much more than the market price of the stock. Before the Company could be classed as either a success or a failure, new explorations had to be undertaken on a depth on the Winona bed and on the lodes in the Baltic horizon approximately 2,500 feet further east.

Although Winona stock was now fully paid, there remained in its treasury 33,333 shares of unsubscribed stock. The long list of some 1,600 stockholders included an unusual number of names well known in business and financial circles. The Co-operation of all these was necessary for the future welfare of the property.

Owing to limited working capital, operations at the Winona mine were

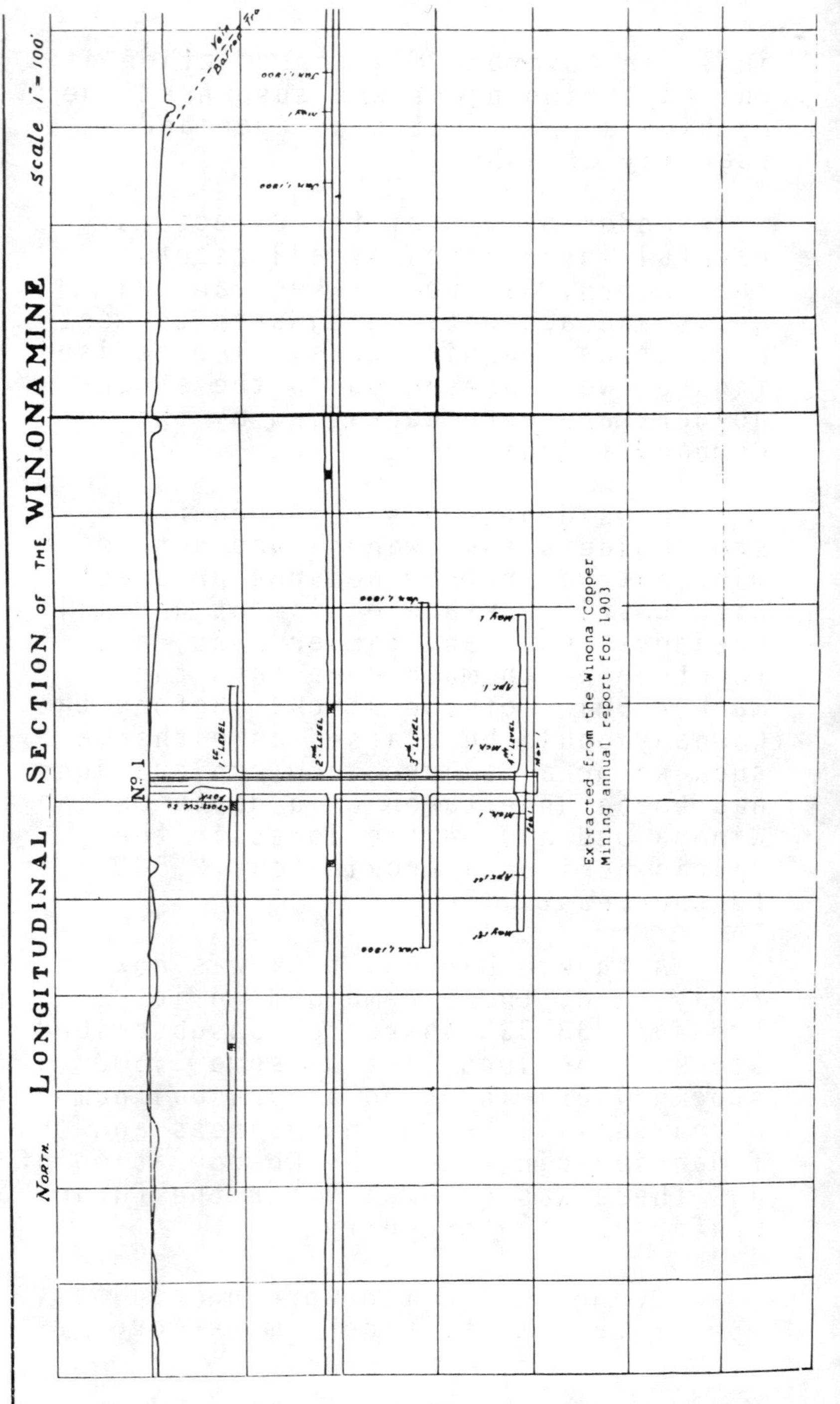

Extracted from the Winona Copper Mining annual report for 1903

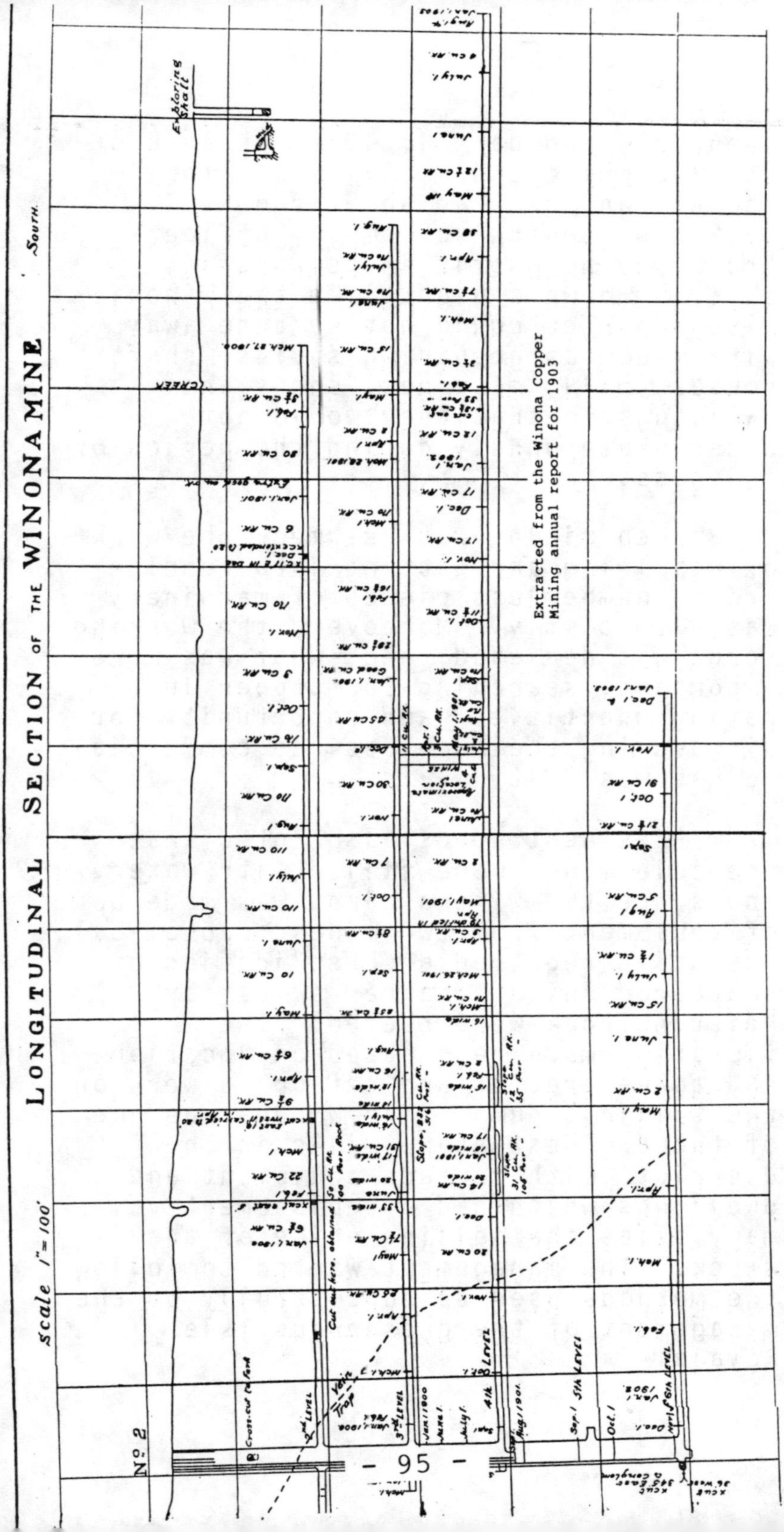

Extracted from the Winona Copper Mining annual report for 1903

suspended February 1, 1919, because of the low price and small demand for copper, and resumed about August 1, 1919. According to Company officers, there was no profit in temporarily shutting down a plant like the Winona. Fixed charges could not be done away with, such as preparing shafts, shaft houses, mill, machines, shops and dwellings so that they would not deteriorate unduly during the period of idleness.

When mining was resumed, the work of repairing and getting into running order, numberless pieces of machinery was also costly. Moreover, the working force was depleted, and, what was more important, searching for copper in paying quantities, and opportunity for discovering the much desired commercial values was lost.

President Woods also noted that the idle mines were filled with water, and since the Winona plant was made up of equipment from two mines, economical operation required a consolidation of scattered units. Larger skips for hoisting rock were needed. The flotation machines housed on location had to be erected and set up to work on the tailings dump. The Winona had one of the two best stamp mills in the district, with surface equipment and dwellings which had a replacement value many times the selling price of the stock. The management wanted to follow the methods used so successfully by the management of the prosperous Isle Royale.

Picture taken in front of the sauna on Neaminen´s farm near Winona. William Bakko, Alex Harma, Elmer Maki and Rudy Ziemer. Courtesy of the Harold H. Heikkinen collection, Atlantic Mine, Michigan

According to Superintendent Braun, during the fore part of January 1919, the No. 1 shaft K. P. hoist drum was lagged, allowing the hoisting rope to overlap and made it possible to hoist from a greater depth. Then hoisting was suspended in this shaft, resuming again on January 9th. All operations at the mine were suspended January 31st and again resumed July 31st, making the total operating six months. The greater part of August was spent making preparations for resuming operations. With a small labor force the water was kept from the mine during the six months of shut-down. A great part of the shut-down period was spent in doing dead work, that is, putting stopes and levels in shape for economical mining, and repairing surface equipment.

The fuel shortage did not affect the Company as far as operations were concerned, but since October 24, they were compelled to team the coal necessasary for No. 3 power plant from the mill stock pile, a distance of 2,000 feet.

President Woods started his twenty-second annual report for the year of 1920, "In May 1920, because of falling prices and high costs of copper, your Directors closed down the mine and mill, placing the property in charge of Captain Sammual Chynoweth, an experienced mining man, as caretaker."

"The extensive and costly plant at Winona, including an efficient mill

September 12, 1992

One of the many poor-rock piles
September 12, 1992

close to the shafts on the Company's own railroad, warranted a search for higher values below 1,800 feet in King Philip ground where the showing is more encouraging, as soon as conditions improved. It would be for the stockholders to decide whether to abandon the proposition or to unite in an effort, at moderate expense, to make a mine of this partially developed prospect." Except for the Treasurer's statement, that was the end of the annual reports.

According to The Copper Deposits of Michigan, the mine's total recorded production from 1902 to 1920 was 1,262,678 tons of rock, yielding 17,684,234 pounds of copper, an average of 14 pounds of copper to the ton of ore. The mineralization was the same type as that of other lodes. Pumpellyite, quartz, epidote, and calcite were plentiful. Prehnite, laumontite, and other minerals were present. The copper was associated with quartz and pumpellyite.

Although the mine was officially closed, the annual report for 1921 informed the stockholders that no mining was done that year and the Boston and Michigan expenses had been kept as low as possible. Some of the Company assets included the mine office dispensary at $1,557.42; Amphidrome Company stock, $400.00; King Philip property, $1,215,334.36; Winona mining property, $450,000.00; real estate, $18,581.33; shafts, etc., $75,255.79; construction, $561,989.83; stamp mill, $426,198.85 and the railroad

$11,742.67. They also had a copper inventory of 130,546 pounds, at 13 cents a pound, for $16,970.98.

During 1921, Winona Directors completed negotiations with the Pampa Land Company for the sale of the timber standing on the property, with Winona retaining all mineral rights. The Pampa Company, having extensive holdings on neighboring sections, was establishing a permanent woodworking plant at the location, which was of benefit to the Winona Company. It created a demand for the Company homes and surplus supplies, helped to support the school, and promised cheap fuel, in the shape of mill-waste, for future mine explorations.

President Woods said that it was his opinion that the cross valleys of the Sleeping River and its branches marked a mineralized shear-zone, under high ground, south of No. 2 King Philip shaft, and the property with its costly surface plants and modern mill, should not be condemned until the short drifts south from No. 2 shaft had been extended for a thousand feet to test this promising ground, or until No. 1 King Philip had been sunk a like distance below the limit at the fifteenth level.

During 1921, the Pampa Company built a sawmill with the Glaven Brothers, Charles, Raymond, and William in charge. They operated all of 1922, with the sawmill burning to the ground one year later. The Brothers rebuilt their mill in the center of town by the power plant and operated until 1929.

President Woods had a bit more to say in his annual report for 1922. They had done a little mining and the explorations in King Philip No. 2 shaft had uncovered on the 8th and 10th levels fine and coarse copper, including a mass of native copper more than 2 feet wide by 6 feet long, the largest ever found in the mine. These discoveries were most encouraging for the reason that it was mass and and barrel copper, disappointingly lacking in the older Winona workings, which now sweetened the product of a successful mine.

Engineer Rex R. Seeber, who for some years had been in charge of the operations at this property, had reported an excellent showing of copper on the 10th level drift. The Company's stamp mill would be put in operation when an adequate tonnage of rock was assured.

The Company pumped water from the shafts all during 1922, but shut off their pumps during the spring of 1923. No mining was completed.

At this time, the combined salaries of the president, secretary-treasurer and engineer-in-charge amounted to $233.00 per month. The business of the Company in Boston was carried on in a small office, half the rent of which was paid by another tenant.

President Wood emphasized that an idle mine, like a deserted house, soon falls into ruin; and believed it was better business to use the assets of

Residents of Winona, Michigan - courtesy of the Harold H. Heikkinen Collection, Atlantic Mine, Mich

the corporation in an effort to find a worth-while copper despoit rather than to let them finally be wasted in paying taxes and the expenses of caretaking. The directors sought and found a buyer for the timber, who had built a mill and provided the Company with cheap fuel. The Winona also sold stamp-sand for road surfacing and further helped to finance Winona's exploration by sales of stock, bid in during previous years at delinquent sales. President Wood asked the stockholders for more money at the end of his statement.

Revenue for the Winona this year consisted of $4,103.88 from the sale of stamp sand; $1,538.00 from rent collections; $686.81 from sale of supplies, and one dividend of 50 cents.

The Winona may have been officialy closed, but activities during 1923 were still being completed. The stockholders were informed that as a result of the exploratory work undertaken and finished during the year 1923, the prospects of making a mine at Winona had been greatly improved.

The hoisting plant at King Philip No. 2, damaged by fire some years before, was repaired and put into operation. This shaft was then freed of water, and a concrete dam built at the 8th level to prevent further flow from the old mine. More important than these accomplishments were new discoveries of copper of commercial grade, and the proving of a continuous copper shoot pitching, at a flat angle, for a distance of 2,000 feet or more, down the lode across No. 2 from the 8th level north to the 12th, south of the shaft. Funds for the operations were provided by sales of mill-tailings and unnecessary machinery.

President Woods wrote "We believe so promising an opportunity finally to realize on an investment of millions is worthy of consideration on the part of our 1,800 stockholders; and that the present showing warrants, at a favorable time, some form of financing for the purpose of putting Winona's costly mine and mill equipment to profitable use."

The annual report for 1924 was short, as President Woods told the stockholders that no work of any kind had been done at the mine since last year. Although he still believed in the mine, he again expressed belief that the showing here warranted, at a favorable time, a reorganization of the corporation, and if possible a merger with a neighboring company, in the effort to put Winona's costly mine and mill equipment to profitable use in large-scale operations such as were now necessary for successful mining in the Michigan district.

The last annual report was for the year of 1925, a two page finance report. They still had assets of $2,693,837.40 and as much in liabilities. This last report was signed by Arthur C. Paine, Treasurer.

The <u>Boston News Bureau</u> published a notice of dissolution of the Winona Copper Company on December 1, 1928. It was on file with the county clerk of

Houghton County, Michigan. The charter of the Company expired by limitation November 7, 1928, without renewal. The charter, it was announced, was not renewed as the stock was fully paid up (par $25.00.) The stock transfer books had been closed by order of the Board of Directors for the purpose of liquidation.

The Saint Mary's Mineral Land Company sold 724 1/100 acres of land on November 13, 1905, for $207,690.00 to the King Philip Copper Company. It was a new Company organized to acquire and and operate 1040 acres of land contiguous to the property of the Winona Copper Company. In payment for the land, the King Philip Company issued, pro rata, to the owners of the land, 50,000 shares of its capital stock, with $6.00 per share endorsed as paid in. The Saint Mary's Mineral Land Company had sold property to the Winona Company in 1898, and the Winona had opened copper ore ground very extensively, with the underground work having been carried on without interruption since the organization of the Company.

The Winona mining people found that the best ground had been located near the dividing line between the Winona and the new King Philip. The third and fourth levels had been carried to that point and discontinued, as they could go no further. On the strength of these developments and the result of St. Mary's own diamond drilling on the King Philip property, it was decided to organize a Company to acquire 724 1/000 acres of St. Mary's land and 320 acres belonging to other

Part of the mining ruins across from the school

September 12, 1992

owners. This was necessary to make a compact body of land, and be able to sink a shaft to a considerable depth. Sinking was begun in December, and the shaft was down some 40 feet in the over-burden, there being about 75 feet of it to penetrate in order to reach the ledge. The plan was to sink the shaft to a depth of about 700 or 800 feet, at the rate of about 60 feet a month, and at that depth to explore the lode thoroughly.

The King Philip Company had been provided with cash by the sale to the St. Mary's Mineral Land Company of 20,000 shares of its capital stock at $6.00 per share. 50,000 shares of its capital stock were issued to pay for its land as already stated, so there remained 30,000 shares in its treasury. From the proceeds of the sale of 20,000 shares the King Philip Company repaid to the St. Mary's Company the sum of $17,084.70 which had been spent in exploring the new King Philip tract.

The Sixth Annual Report of the Saint Mary's Mineral Land Company for 1906 included a separate annual report for the King Philip Copper Company. The officers for the King Philip serving during 1906 listed Arthur G. Stanwood as President, George P. Gardner as Vice-President and Edward B. O'Connor as Secretary and Treasurer. The Board of Directors consisted of Nathaniel Thayer, George P. Gardner, Charles J. Paine, Jr., Walter Hunnewell, Nathaniel H. Stone, Arthur G. Stanwood and Rufus R. Goodell. The Company's office was located at 713 Sears Building, Boston, Mass.

The Winona Mining Company's Stamp Mill as seen in the early 1900's. Courtesy of the Roy W. Drier Collection, Library Archives, Michigan Technological University

During 1906, the King Philip Copper Company spent $300,000.00 for real estate at the mine location. $96,076.97 was expended for such things as shaft sinking, pumps and construction, exploration, surface, office and engineering expenses, boiler house construction, a new compressor, hoist, employee houses, camp construction and equipment, a blacksmith shop and a new shaft house. Also included was a derrick, office, dry house, drainage system, water supply, timber and other miscellaneous items.

Since the new company's organization in November 1905, not much work had been possible until the ground was free from snow in the spring of 1906. Then an inclined shaft was started in the trap rock underlying the Winona lode. It was expected that this shaft would reach a depth of 400 feet about April 1, 1907, when a cross-cut would be made to the Winona lode. At about 30 feet from the surface, a fine quicksand was encountered, and soundings from the bottom of the shaft showed that the supposed depth of the overburden had been underestimated.

By an exchange of about five acres of mineral land with the Winona Copper Company, the boundary line between the two Companies had been made to run at right angles with the outcrop of the lode.

Much construction was completed during 1906. An engine and boiler house, 30 by 40 feet, was built of corrugated iron; a new engine and boiler house was also built in the back of the shaft, which were 24 by 48 feet, and 22 by 24 feet, both with corrugated iron roofs. The blacksmith shop was 18 by 24 feet, of wood-frame construction, covered with tarred paper; a change house 20 by 30 feet and also covered with tarred paper; and a sleeping house 22 by 40 feet, covered with tarred paper.

Two wood-frame boarding houses, one 15 by 18 feet and the other 20 by 40 feet, were covered with tarred paper. The ice house was 16 by 18 feet and the office and warehouse was 22 by 36 feet, both of wood-frame construction and covered with tarred paper. For the employees, 12 wood-frame dwelling houses were built, 20 1/2 by 26 feet, with additions of 4 1/2 by 10 feet, with stone foundations and cellars.

A Nordberg hoist, good for 1,500 feet, was purchased and installed in back of the shaft. The compressor and boiler were also being moved to a wood-frame building that had been erected to house this equipment. A temporary shaft house, 20 by 26 feet and 20 by 20 feet was now in use, but was not yet completed. A poor-rock trestle was being built and would soon be completed. The pumping station at the Sleeping River was 22 by 30 feet and a pipe line ran across the King Philip location to the Winona water tank, furnishing water for the steam plant and the employee's homes. According to General Manager Lucius L.

Hubbard, they expected 1907, the next year, to be a great year since they were to sink a second shaft during that summer and would soon have copper values equal to those of the Winona.

The second annual report for the King Philip Copper Company for the year 1907 showed that President Arthur G. Standwood died during the year. Mr. George P. Gardner was the new President. Expenditures of $121,134.60 showed the usual copper mining expenses but included construction of more employees homes, power house construction and equipment, electric light and telephone construction, another shaft house, coal trestle construction, an air line construction, a new dry, railroad construction and a small amount for the water supply.

The year's end of 1907 showed that shaft number 1 was sunk a total depth of 611 feet. Shaft number 2 was located 2,780 feet southwest of the No. 1 shaft, on the west end of the Sleeping River, above the dam which was begun during September 1907. It was fully timbered and equipped with a temporary shaft house. The sinking was done with a small electric hoist which had been replaced by a steam hoist of larger capacity. During the year the principal construction work was the erection of the compressor plant on the Sleeping River, at about the center of the property, and the erection of ten new employee's dwelling houses. The building for the compressor and boilers was completed, the smoke stack was

SUPERINTENDENT Rex Seeber resided in this grandiose establishment in Winona when he was heading the copper firm there. He later was on the staff of Michigan Tech and during the latter days of his life resided in the former Harris residence in East Hancock.

partly erected, and the greater part of the 70-drill compressor equipment delivered.

A rock-crusher for the No. 1 shaft was delivered and installed in the temporary rock house, so that rock could be shipped to the mill direct from the drifts. Construction during 1907 included the power plant which was 52 by 108, feet combined boiler and compressor house, 52 by 52 feet and 52 by 56 feet respectively. The building had three 13-inch brick walls and one temporary steel wall; also a steel roof and gables.

The steel smoke stack was 165 feet high, 7 feet 6 inches in diameter, was self supporting and brick lined. The new coal trestle was 165.5 feet in length and had a capacity of 2,000 tons of coal. The trestle was provided with a six by six foot concrete-lined adit, inwhich tram cars were loaded by gravity. An air-pipe line was laid of 1,500 lineal feet of 5-inch wrought iron pipe from the Winona shaft No. 4 to the King Philip shaft No. 1. The Company also laid 3,000 feet of 3-inch temporary pipe from King Philip No. 1 shaft to the King Philip No. 2 shaft.

Workers also did a lot of grubbing and excavating for 2,150 feet of permanent 8-inch pipe from the Sleeping River water plant to the No. 1 shaft. All of this pipe had been lying on the ground, causing many problems during the winter months. The Company strung 1,150 feet of electrical- transmission wire for the No. 1 shaft electrical hoist. They also constructed 550 feet of lineal pole line.

For the workers, the company erected ten new six-room single family dwellings. They were wood-frame structures, 20 by 26 feet, with 4 by 10 foot extensions on the rear of each home for a pantry and vestibule.

The King Philip Copper Company's third annual report, for the year of 1908, showed Charles J. Paine Jr as the President of this company effective June 22, 1908. Of course these men were also the same people on many other boards, such as the Saint Mary's Mineral Land Company, the Challenge Location, the Champion Copper Company, the Hancock Consolidated Mining Company, the La Salle Copper Company, the Pacific Copper Company, Copper Range Consolidated Company, Winona Copper Company, Mayflower Mining Company, North Lake Mining Company, and several others.

President Paine also spoke of the Directors having considered the question of building a stamp mill in conjunction with the Winona Copper Company. They had recently appointed a committee with power to join with the Winona Copper Company in building a two-head stamp mill. By re-using the water from the Sleeping River, the Company believed that they would have enough water for a four-head mill, capable of stamping 3,000 tons of rock daily. This opinion was founded on stream measurements made during the course of the past year or two, and on the experience of other mining companies.

General Manager Hubbard reported that they were showing some shot copper and were awaiting the arrival of the new hoist. During 1908, the changes which began during 1907 at the No. 1 shaft were completed and a spur-track from this shaft to the main line was graded by the Copper Range Railroad. The crusher was installed and the head-sheave raised. A motor driven generator from the No. 4 Winona shaft was also connected to the No. 1 shaft hoist and was working satisfactorily.

At the No. 2 shaft a small boiler house was built, and a dry constructed to replace the one that had burned. A temporary shaft house was erected to serve during the development work at this shaft. A five-inch air line was laid between the receiver at the Sleeping River plant and Shaft No. 2, and an eight-inch air line was laid from the former point to the No. 1 shaft. A culvert was also built across the Sleeping River and the waste rock from the No. 2 shaft was being dumped to form a road bed over the stream at this point. Finally, the power plant at the Sleeping River was completed.

The Saint Mary's Mineral Land Company reported that during the year 1909, development work at the King Philip Copper Company, in which the St. Mary's owned 83,905 shares out of 100,000 outstanding, was progressing satisfactorily. The stockholders had already been assessed four times.

The No. 1 shaft was sunk to the 12th level and the hoist at this shaft had about reached the limit of its

The King Philip Shaft No. 2, as seen on Octoɔer 9, 1988

October 9, 1988

capacity. Nine drills were in operation in this shaft. The No. 2 shaft was at its 8th level and a drift was started at that level towards the No. 1.

Much money was spent in construction this year. The power plant had problems during the winter of 1908-1909, as surface water and frost had undermined one of the walls of the compressor plant at the Sleeping River and parts of the floors. This plant was not in commission, and when the damage was noticed, considerable expenditures were necessary for repairs. A 12 and 12 by 20 foot duplex geared Nordberg hoist, good for 1,200 feet and a two-ton skip were installed in the No. 2 hoisting plant.

At the change house, a stone and brick "dry," 38 feet by 92 feet inside, with steel tressed roof and adequate to accommodate 200 men, the Mining Captain and the shift bosses, was built. It contained eight shower baths, steel lockers, steel drying racks and was furnished with steam heating and hot water plants. It was located near the No. 1 hoisting plant.

The original engine and boiler house at the No. 1 shaft was remodeled into a double dwelling house, 28 by 40 feet, each side containing eight rooms, piped for water and wired for electric lights. Also built were twelve single frame seven-room houses, 20 feet by 30 feet, and two eight-room houses, 24 by 36 feet, all piped for water and wired for electric lights. As far as the water supply, 1,150 feet of permanent water lines had been laid on the

The King Philip Shaft No. 2, as seen on October 15, 1988

October 15, 1988

location during the year.

The excavation for foundations of the two-head stamp mill started on the 19th of May 1909. They began to sink the caissons for the stamp heads in July. A railroad track was graded and laid from the main spur to the rear of the mill, for delivery of rock, sand, cement and other material. The excavation for the concrete core at the dam was also furnished and the laying of the core begun. Early in September, both caissons were bottomed at about 57 feet on a stratum of clay, and the caissons were filled with concrete.

Early in December of 1909, all the foundations for the mill and coal trestle were completed and the boiler house and smoke stack erected. By January 1, 1910, erection of steel over the lower end of the mill had advanced considerably, and the steel coal trestle was finished. The electric locomotive had been in service for some weeks. Some delays had occurred in the delivery of steel, which delayed the completion. Parts of the machinery had been delivered and much of the remainder was ready for delivery, so there wouldn't be any other delays.

For the second year (1910) the Saint Mary's Mineral Land Company had been marked by several large sales of land. These included property sold to the Houghton Copper Company, the Franklin Mining Company and the Quincy Mining Company.

King Philip's fifth annual report showed $138,000.00 being advanced to

THIS is the King Phillip Location of the Winona Copper Co. in Elm Township, near the Ontonagon County borderline. The King Phillip site currently is the most active of the overall Winona mining setup.

building the stamp mill. President Charles J. Paine Jr. said "During the past year the developments on your property have been as encouraging as during any previous year. The ground showed a fairly encouraging amount of copper, and the first head of the stamp mill that is being built by the Company in conjunction with the Winona Copper Company will be finished very shortly. The second head will be completed as soon as it will be needed."

General Manager Lucius L. Hubbard mentioned in his report that the tributary to the No. 1 shaft had more than enough good ground opened to supply all the rock that could be handled in their temporary rock house. Before the mine could be worked to its full capacity, it was likely that more houses would be needed for employees. No special construction had been carried on during the year except the building of the stamp mill.

The _Copper Handbook_ reported that by 1910 the mine was opened by two shafts, both sunk in the footwall, a wise precaution, and the mine was connected with the Winona on the 5th, 8th and 10th levels. Rock filling was being used underground. The mine ended 1910 working eleven drills. The No. 2 shaft had an isolated steam hoisting plant, and both shaft houses were temporary. The King Philip took electrical power from the Winona and in return furnished compressed air. The coal for the boiler-house was drawn through a concrete tunnel, under a coal-trestle surmounted by a railroad track.

According to the Copper Handbook, buildings included a warehouse, smithy, changing house, office, cookhouse, bunkhouse and about 40 dwellings for employees, these being six-room electric-lighted homes, thoroughly comfortable and homelike.

The eleventh annual report of the Saint Mary's Mineral Land Company for the year 1911, is quite thin. President George P. Gardner reported that "The King Philip Copper Company was sold to the Winona Copper Company in May. Each three shares of King Philip Copper Company stock received two shares of Winona Copper Company stock. The usual King Philip report was absent.

The last attempt to mine copper from the Winona and King Philip started in 1968 by the Lake Superior Copper Company. It was a closed corporation and had no plans to put stock on the market. Their efforts took place at the King Philip No. 2 shaft, on the southwest part of the property, near the Houghton and Ontonagon border. The opening was 2,700 feet south of the King Philip No. 1 and on the south side of the Sleeping River. Both No. 1 and No. 2 shafts were connected on the eighth level.

January of 1969 found the King Philip property being dewatered and all efforts were concerned with dewatering the overall Winona mine shafts. At that time no actual mining had taken place and there had been only a slight amount of cleanup work done. Actually there was no need for this type of endeavor inasmuch as the levels were free of any form of fallen rock or other debris.

The company had been performing development work in the former King Philip shaft with the result that tests were made of the rock and all had been entirely favorable. There were about one and one-half million tons of ore in the stopes and it was hoped to hoist this to the surface within the next year or two.

When the property closed, the miners left most of the equipment they were using in the shaft, having always operated under the theory that the property would resume active operations. Because of this there were many mining cars, much rail, tools and the like still in the mine and frequently these were encountered by the mine workers as they moved along in their development work.

Some of the equipment had been hoisted to the surface but the greater amount had been permitted to remain in the shaft for the purpose of being analyzed. King Philip No. 2 was 1,108 feet deep in 1917 when it was closed. There were crosscuts on the sixth, eight and tenth levels to the Winona.

Part of the mining unit consisted of Joseph Plesha of Painesdale, Elvin Lindeman of Winona and Henry Mickalich of South Range. James Riipa was a watchman and John Katalin was the hoist engineer.

It was indicated that this current company was not concerned with actual mining, but only exploration and factors involving determination of prospects. After a short period, the Lake Superior management found that Mother Nature did not pay off in a metallic way that they had envisioned, and that Company also quit the Winona - King Philip propertys.

SOURCES

The references listed below were used in gathering information to aid in the writing of this publication. Not all of the sources are listed however, as many people of the Copper Country area provided much information.

I particularly used the resources of the Michigan Technological University Library and the Daily Mining Gazette, both located in Houghton. I am indebted to Theresa S. Spence and Verna K. Masters of the Michigan Technological University for their assistance from the Library Archives.

PUBLICATIONS

Mining in Michigan, A Catalog of Company Publications, 1845-1980, by LeRoy Barnett, 1983

Annual Reports for the Winona Copper Company, 1899 through 1925

Prospectus of the Winona Mining Company of Michigan, 1864

Annual Reports for the Saint Mary's Mineral Land Co., 1905 through 1911

The Copper Deposits of Michigan, Professional Paper 144, by B. S. Butler and W. S. Burbank, 1929

The History of Winona, by Patti Leppanen, 1984

Tests of an Ilgner Electrical Hoist, published in the Transactions of American Institute of Mining Engineers, by Rex, R. Seeber, Winona, MI, 1910

Cemeteries of the Copper Country, by David H. Thomas, 1973

NEWSPAPERS

The Daily Mining Gazette, Houghton, Michigan

The Native Copper Times, Lake Linden, Michigan

The Evening Journal, Hancock, Michigan

The Keweenaw Miner, Mohawk, Michigan

The Miner's Bulletin, Hancock, Michigan

The Calumet News, Calumet, Michigan

OTHER

Micro-copy of Record of Appointment of Postmasters, 1832 - Sep 30, 1971, National Archives

Interview with William I. Siltala, of Atlantic Mine

Technical advice and assistance from Edward Charles, Adams Township Supervisor, of South Range

Assistance from Tauno Kilpela, of Atlantic Mine

Information from Jeanne Gourneau, of Twin Lakes

Telephone interview with Flora M. Knudsen, of Winona

Interview with Uno Sarja of Winona

The following folders were used from MTU Archives' vertical files:

Copper Mines and Mining - Companies, Winona Copper Company

Schools, Winona

Cities and Towns - Winona

Copper Mines and Mining Companies, Winona

Winona - clippings

Near Winona

Winona Mining Company

Winona Copper Company

Churches - Winona

ADD THIS COPPER COUNTRY LOCAL HISTORY SERIES TO YOUR PERSONAL LIBRARY

COR-AGO, A LAKE LINDEN MEDICINE COMPANY
First of a local history series

A COPPER COUNTRY LOGGER'S TALE
Second of a local history series

GREGORYVILLE - THE HISTORY OF A HAMLET LOCATED ACROSS FROM LAKE LINDEN, MICHIGAN
Third of a local history series

WHITE CITY - THE HISTORY OF AN EARLY COPPER COUNTRY RECREATIONAL AREA
Fourth of a local history series

SOME COPPER COUNTRY NAMES AND PLACES
Fifth of a local history series

THE HISTORY OF LAKE LINDEN, MICHIGAN
Sixth of a local history series

THE HISTORY OF JACOBSVILLE AND IT S SANDSTONE QUARRIES
Seventh of a local history series

THE HISTORY OF COPPER HARBOR, MICHIGAN
Eight of a local history series

THE HISTORY OF EAGLE HARBOR, MICHIGAN
Ninth of a local history series

LAKE LINDEN'S YESTERDAY - A PICTORIAL HISTORY, VOLUME I
Tenth of a local history series

THE HISTORY OF EAGLE RIVER, MICHIGAN
Eleventh of a local history series

JOSEPH BOSCH AND THE BOSCH BREWING COMPANY
Twelfth of a local history series

COPPER FALLS - JUST A MEMORY
Thirteenth of a local history series

THE CALUMET THEATRE
Fourtheenth of a local history series

EARLY DAYS IN MOHAWK, MICHIGAN
Fifteenth of a local history series

LAKE LINDEN'S YESTERDAY - A PICTORIAL HISTORY, VOLUME II
Sixteenth of a local history series

THE KEWEENAW WATERWAY
Seventeenth of a local history series

A BRIEF HISTORY OF AHMEEK, MICHIGAN
Eighteenth of a local history series

ALL ABOUT MANDAN, MICHIGAN
Nineteenth of a local history series

HANCOCK, MICHIGAN, REMEMBERED, VOLUME I
Twentieth of a local history series

THE SETTLING OF COPPER CITY, MICHIGAN
Twenty-first of a local history series

LAKE LINDEN'S YESTERDAY - A PICTORIAL HISTORY VOLUME III
Twenty-second of a local history series

PAINESDALE, MICHIGAN - OLD AND NEW
Twenty-third of a local history series

SOME OF THE BEST FROM C & H NEWS - VIEWS VOLUME I
Twenty-fourth of a local history series

HANCOCK, MICHIGAN, REMEMBERED - CHURCHES OF HANCOCK, VOLUME II
Twenty-fifth of a local history series

OJIBWAY, MICHIGAN, A FORGOTTEN VILLAGE
Twenty-sixth of a local history series

LAURIUM, MICHIGAN'S EARLY DAYS
Twenty-seventh of a local history series

DELAWARE, MICHIGAN, ITS HISTORY
Twenty-eight of a local history series

LAKE LINDEN'S LIVING HISTORY - 1985
Twenty-ninth of a local history series

SOME OF THE BEST FROM C & H NEWS - VIEWS VOLUME II
Thirtieth of a local history series

THE GAY, MICHIGAN, STORY
Thirty-first of a local history series

EARLY RED JACKET AND CALUMET IN PICTURES
Thirty-second of a local history series

LAKE LINDEN'S DISASTROUS FIRE OF 1887
Thirty-third of a local history series

PHOENIX, MICHIGAN'S HISTORY
Thirty-fourth of a local history series

FREDA, MICHIGAN, END OF THE ROAD
Thirty-fifth of a local history series

HOUGHTON IN PICTURES
Thirty-sixth of a local history series

THE COPPER RANGE RAILROAD
Thirty-seventh of a local history series

LAC LA BELLE
Thirty-eight of a local history series

TRIMOUNTAIN AND ITS COPPER MINES
Thirty-ninth of a local history series

EARLY RED JACKET AND CALUMET IN PICTURES, VOLUME II
Fortieth of a local history series

UPPER PENINSULA'S WOLVERINE
Forty-first of a local history series

REDRIDGE AND ITS STEEL DAM
Forty-second of a local history series

THE MINERAL RANGE RAILROAD
Forty-third of a local history series

WINONA AND THE KING PHILIP PROPERTIES
Forty-fourth of a local history series

A BRIEF LIST OF PUBLICATIONS PERTAINING TO COPPER COUNTRY HISTORY
(Not part of the series)